Accessible Public Transportation

The United States is home to more than 54 million people with disabilities. This book looks at public transit and transportation systems with a focus on new and emerging needs for individuals with disabilities, including the elderly. The book covers the various technologies, policies, and programs that researchers and transportation stakeholders are exploring or putting into place. Examples of innovations are provided, with close attention to inclusive solutions that serve the needs of all transportation users.

Aaron Steinfeld is an Associate Research Professor in the Robotics Institute at Carnegie Mellon and the Co-Director of the Rehabilitation Engineering Research Center on Accessible Public Transportation. He earned his PhD, MS, and BS in Industrial and Operations Engineering from the University of Michigan and completed a postdoctoral position at the University of California, Berkeley. His interests focus on constrained user interfaces and operator assistance, predominantly in the realms of human-robot interaction, transportation, and intelligent systems.

Jordana L. Maisel is the Director of Research Activities at the IDeA Center at the University at Buffalo (UB), and an Assistant Research Professor in the School of Architecture and Urban and Regional Planning at UB. She earned her BS in Human Development from Cornell University, and her Masters of Urban Planning and PhD in Industrial and Systems Engineering from UB. She currently serves as a Project Director in the Rehabilitation Engineering Research Center on Universal Design and the Built Environment and a Project Co-Director in the Rehabilitation Engineering Research Center on Accessible Public Transportation.

Edward Steinfeld is a State University of New York Distinguished Professor of Architecture at the University at Buffalo. He received his Bachelor of Architecture degree from Carnegie Mellon University and an MArch and ArchD from the University of Michigan. He is the founding Director of the Center for Inclusive Design and Environmental Access (IDeA), a center of excellence in universal design. He is the Co-Director of the Rehabilitation Engineering Research Center on Accessible Public Transportation and the Director of the Rehabilitation Engineering Research Center on Universal Design and the Built Environment.

Written by leading researchers in the field, *Accessible Public Transportation* is a thoroughly readable primer for everyone—policymakers, planners, and advocates new to the topic. Synthesizing traditional and emerging approaches to public transit for people with disabilities, the authors offer research findings and everyday practices that are informative and can be adapted to our own communities.

—Joey Goldman, Nelson\Nygaard Consulting,
Co-Chair, TRB Committee on Accessible Transportation and Mobility

Accessible Public Transportation

Designing Service for Riders with Disabilities

Edited by Aaron Steinfeld, Jordana L. Maisel, and Edward Steinfeld

Routledge
Taylor & Francis Group

NEW YORK AND LONDON

First published 2018
by Routledge
605 Third Avenue, New York, NY 10017

and by Routledge
2 Park Square, Milton Park, Abingdon, Oxon, OX14 4RN

First issued in paperback 2021

Routledge is an imprint of the Taylor & Francis Group, an informa business

© 2018 Taylor & Francis

The right of Aaron Steinfeld, Jordana L. Maisel, and Edward Steinfeld to be identified as the authors of the editorial material, and of the authors for their individual chapters, has been asserted in accordance with sections 77 and 78 of the Copyright, Designs and Patents Act 1988.

Publisher's Note
The publisher has gone to great lengths to ensure the quality of this reprint but points out that some imperfections in the original copies may be apparent.

Library of Congress Cataloging-in-Publication Data
Names: Steinfeld, Aaron, author. | Maisel, Jordana, author. | Steinfeld, Edward, author.
Title: Accessible public transportation: designing service for riders with disabilities / Aaron Steinfeld, Jordana L. Maisel, and Edward Steinfeld.
Description: 1 Edition. | New York: Routledge, 2018. | Includes bibliographical references and index.
Identifiers: LCCN 2017013895 (print) | LCCN 2017029120 (ebook) | ISBN 9781315118321 (Master) | ISBN 9781482234114 (WebPDF) | ISBN 9781351644600 (ePub) | ISBN 9781351635110 (Mobipocket/Kindle) | ISBN 9781482234107 (hardback: alk. paper) | ISBN 9781315118321 (ebk) Subjects: LCSH: People with disabilities—Transportation. | Transportation—Barrier—free design. Classification: LCC HV3022 (ebook) | LCC HV3022 .S74 2018 (print) | DDC 362.4/0483–dc23
LC record available at https://lccn.loc.gov/2017013895

ISBN 13: 978-1-03-224210-1 (pbk)
ISBN 13: 978-1-4822-3410-7 (hbk)

DOI: 10.4324/9781315118321

Typeset in Times New Roman
by codeMantra

Contents

List of Illustrations   vii
List of Contributors   ix
Preface   xiii
Acknowledgments   xvii

1 **The Importance of Public Transportation**   1
   AARON STEINFELD AND EDWARD STEINFELD

2 **The Culture of Accessible Transportation**   5
   EDWARD STEINFELD

3 **The Scope of Inclusive Transportation**   20
   EDWARD STEINFELD

4 **Trip Planning and Rider Information**   32
   AARON STEINFELD, HEAMCHAND SUBRYAN, AND ELLEN AYOOB

5 **The Built Environment**   39
   EDWARD STEINFELD

6 **Vehicle Design**   54
   JAMES LENKER, CLIVE D'SOUZA, AND VICTOR PAQUET

7 **Demand Responsive Transportation**   68
   EDWARD STEINFELD AND AARON STEINFELD

8 **Paratransit Scheduling and Routing**   83
   ZACHARY B. RUBINSTEIN AND STEPHEN F. SMITH

9 **Location-Based Information**   90
   AARON STEINFELD, ANTHONY TOMASIC, YUN HUANG,
   AND EDWARD STEINFELD

10 Social Computing and Service Design 98

JOHN ZIMMERMAN, AARON STEINFELD, AND ANTHONY TOMASIC

11 Learning from Riders 106

JORDANA L. MAISEL, DAISY YOO, JOHN ZIMMERMAN,
EDWARD STEINFELD, AND AARON STEINFELD

12 Vision for the Future 114

AARON STEINFELD AND EDWARD STEINFELD

References 123
Index 137

List of Illustrations

Figures

2.1 Bus shelter in New York City awarded through a franchise-bid program 11
2.2 Flyaway suburban bus terminal with nearby parking facility
(a) main entry to the terminal, (b) large parking structure on site 12
2.3 SEPTA training facility for riders (a) simulated bus, (b) simulated
subway station platform 12
2.4 Plan view of Transect diagram based on the SmartCode 17
2.5 The Travel Chain 18
2.6 Full-scale simulation of existing low-floor bus 19
2.7 Image of virtual model of the simulated bus 19
3.1 Photo of fare gates in a major city's subway system. The wider
"accessible" fare gate proved very popular with all riders 23
3.2 BRT system in Rio de Janeiro, Brazil. (a) Bus and dedicated
right-of-way, (b) boarding platform 31
4.1 A multi-language information display inside a Japanese subway car 36
5.1 BRT line in Los Angeles, CA, showing relationship to street, park,
and bike path 41
5.2 Although this bus stop is accessible by ramp, the entry to the shelter
is too narrow to be accessible for wheelchair users and the shelter
also blocks access to most of the stop. A bus would have to pull up
in front of the shelter to load a wheelchair user 42
5.3 This windscreen at a large bus shelter in a cold climate provides
protection on all sides. Two entries improve access and security. The
entries and the interior are wide enough for wheelchair access. The
tinted glass reduces heat gain in summer and increases visibility of
the stop from a distance 43
5.4 Diagrams of (a) "tree-like" and (b) "ring-like" designs 44
5.5 Real-time information provided on large, easy-to-read
dynamic sign display 45
5.6 BART tactile map 46
5.7 Multisensory Interactive Touch Model of a campus that provides
tactile and audible feedback as the user touches different
buildings and streets 48
5.8 An elevated commuter rail station in the SEPTA system was
equipped with a long ramp rather than an elevator based on the
preferences of riders with disabilities who were consulted in its design 49

5.9   Solutions for accommodating horizontal and vertical gaps:
(a) portable ramp and (b) folding ramp   50
5.10 Image of DC Metro platform with embedded lighting   52
6.1   (a) Lifts used in traditional high-floor buses to allow passengers
who cannot board or exit via stairs, but are costly to maintain and
increase bus dwell times at local stops, (b) ramps used in low-floor
buses are less expensive to maintain and generally allow more
efficient boarding and exiting of the bus   56
6.2   Steeper ramp slopes such as those encountered when a ramp is
deployed to the roadside present more usability problems than
shallower ramp slopes that result when a ramp is instead deployed to
sidewalk or bulb out   58
6.3   Plan views of the front half of three interior bus configurations,
along with mannequin placement for simulated conditions of low
and high passenger loading (D'Souza et al., 2017). (a) Front entry-
forward exit configuration with forward-facing seats, (b) rear
entry-forward exit with side-facing seats, (c) rear entry-rear exit with
forward-facing seats   60
6.4   Ramp deployment for an entry-exit located near the middle of the
bus allows passengers more room when maneuvering inside the bus   62
6.5   Electronic fare reader reduces fare payment time but when combined
with a cash payment option requires a safe below, thus reducing
clear floor space at the bus entry   63
6.6   Traditional and alternative wheelchair securement systems   65
7.1   The MV1, an accessible taxi vehicle   72
7.2   Accessible Uber vehicle   73
7.3   Shuttle buses. (a) High-floor shuttle bus with lift, (b) high-floor
shuttle bus with stairs, and (c) low-floor shuttle bus with ramp   75
7.4   (a) Bridj vehicle, (b) service area map   75
7.5   Google autonomous vehicle. This vehicle is too small for many
people to exit and enter   77
7.6   Exterior of EZ10 self-driving transport vehicle with ramp deployed   78
9.1   Smartpen used with tactile map   92
10.1 Tiramisu interface screens. (a) Main map, (b) select route,
(c) before recording, (d) during recording, (e) agency report
categories, (f) filing a report   103
11.1 Results from the Guided Tours study. Questions Asked Relating to
Ease of Task: (bold indicates responses are graphed above). How
difficult was it to … **(5) plan your trip**; (8) find information about the
routes; (11) understand the information you found about the routes;
(15) understand the schedule and route maps; **(32) to select the correct
ticket to purchase**; **(35) complete the ticket purchase**; (55) get off the
train; **(62) use the schedule and route map to find a bus stop**; **(65) find
the bus stop**; (68) use the bus stop's shelter; (80) pay the bus fare?   109
11.2 Conditions affecting routine activities   111
11.3 Suggestions for bus service of the future   112

Table

11.1 Top ten reported problems experienced by transit bus riders   111

List of Contributors

Ellen Ayoob MDes is an Adjunct in the Human Computer Interaction Institute and most recently was the Director of Education at the Creative Nonfiction Foundation, a small nonprofit dedicated to the art of telling true stories. Before that she was a design researcher in the Human Computer Interaction Institute and the Robotics Institute at Carnegie Mellon University. She has an MDes from the School of Design at Carnegie Mellon University and has spent much of her career at the intersections of human-centered design, engineering, and writing.

Clive D'Souza PhD is Assistant Professor in the Department of Industrial and Operations Engineering at the University of Michigan, Ann Arbor. He directs the Inclusive Mobility Laboratory within the U-M Center for Ergonomics and is a faculty affiliate at the U-M Center for Occupational Health and Safety Engineering. He joined the University of Michigan after earning a PhD in Industrial Engineering and an MS in Mechanical Engineering from the University at Buffalo. During this time, he was a Research Assistant on the Rehabilitation Engineering Research Center (RERC) on Accessible Public Transportation. He currently leads a National Institute on Disability, Independent Living, and Rehabilitation Research (NIDILRR)-funded field-initiated project on measuring inclusivity in public transportation systems. His research and teaching interests focus on ergonomics and human factors to address human performance and universal design in occupational, health care, and transportation settings, and he has published multiple peer-reviewed articles on these topics.

Yun Huang PhD is Assistant Professor at the iSchool in Syracuse University. She co-directs the SALT (Social Computing Systems) lab (http://salt.ischool.syr.edu). Before joining the iSchool, she was a Postdoctoral Fellow of the Robotics Institute at Carnegie Mellon University. She received both her master's degree and doctorate from the Donald Bren School of Information and Computer Sciences (ICS) at UC Irvine. She earned her bachelor's degree from the Department of Computer Science and Technology at Tsinghua University, Beijing, China. Her research examines new context-driven ways of designing mobile social systems, particularly focusing on crowdsourcing systems. She employs HCI research methods to understand how people use mobile social systems to co-create new services and to co-perform tasks by contributing observations of the environment, by sharing information about themselves or by accomplishing some tasks that require labor or human intelligence.

James Lenker PhD, OTR/L is Associate Professor in the Department of Rehabilitation Science at the University of Buffalo (UB). He is a licensed occupational therapist with more than 25 years of experience in the assistive technology (AT) field. He is noted for his expertise in AT outcomes measurement, which has included co-Investigator roles on two NIDILRR-funded Disability and Rehabilitation Research and Programs (DRRPs) focusing on AT outcomes research. He serves as a Project Co-Director for Rehabilitation Engineering Research Center on Universal Design and the Built Environment (RERC-UD) and Rehabilitation Engineering Research Center on Accessible Public Transportation (RERC-APT) research projects. He has applied numerous user-centered research methods in his research including real-time observer ratings of task performance, self-reported user usability ratings, semi-structured interviews, online surveys and focus groups. He is an active member of the Rehabilitation Engineering & Assistive Technology Society of North America (RESNA) and has served three terms on its Board of Directors and Executive Committee. He is currently an Associate Editor for *Assistive Technology.*

Jordana L. Maisel PhD is Director of Research Activities at the Center for Inclusive Design and Environmental Access (IDeA), which is housed within the School of Architecture and Urban and Regional Planning at the University at Buffalo (UB). She served as a Co-Director of the RERC-UD, and currently serves as a Project Director in the RERC-UD and the RERC-APT. She is also Assistant Research Professor in the School of Architecture and Urban and Regional Planning at UB. She has lectured at numerous conferences across the country and has written many peer-reviewed articles. She co-authored the first textbook on universal design, *Universal Design: Creating Inclusive Environments* (Wiley & Sons, Inc., 2012) and is co-editor of the forthcoming *Inclusive Design* (Routledge, 2017).

Victor Paquet ScD is Associate Professor of Industrial & Systems Engineering at UB. He is Co-Director of the UB's Occupational Safety and Health Training Grant Program and UB's Center for Excellence in Home Health and Well-Being through Adaptive Smart Environments (Home-BASE). He was Co-Director of the Anthropometry of Wheeled Mobility research program for the U.S. Access Board and Co-Leads key projects in the RERC-UD and the RERC-APT. He has authored or co-authored over 95 peer-reviewed journal articles and conference publications in the areas of universal design and occupational ergonomics. In 2004, he edited a Special Issue of the *International Journal of Industrial Ergonomics* on "Anthropometry and Disability," and in 2010 he co-edited a Special Issue of *Assistive Technology* on "Space Requirements for Wheeled Mobility."

Zachary B. Rubinstein PhD is Principal Project Scientist in the Robotics Institute at Carnegie Mellon University. He earned his PhD in 2002 from the University of Massachusetts at Amherst and, prior to coming to Carnegie Mellon University, he was Assistant Professor in Computer Science at the University of New Hampshire. Through his work in both industry and academia, his research has focused on developing intelligent solutions to real-world problems in the area of effective resource allocation in dynamic and uncertain environments. He has been a Principal Investigator and Co-Principal Investigator on a number of both government and private sponsored research projects. Over the last several years he has focused on transportation-based technologies, developing real-time optimization and management systems for paratransit and home health care services, and contributing to the development of an advanced adaptive traffic light signalization system. He has co-authored over 30 published articles in the areas of artificial intelligence, automated planning and scheduling, multi-agent systems, case-based reasoning, and blackboard systems.

Stephen F. Smith PhD is Research Professor of Robotics and Director of the Intelligent Coordination and Logistics Laboratory at Carnegie Mellon University. His research focuses broadly on the theory and practice of next-generation technologies for automated planning, scheduling, and coordination. He pioneered the development and use of constraint-based search and optimization models, and has successfully fielded artificial intelligence-based planning and scheduling systems in a range of application domains. In recent years, one of his principal application interests has been that of reducing congestion and improving mobility in urban environments. This has led to development and deployment of new technologies for dynamic, same-day paratransit scheduling and real-time urban traffic signal control, and his current work aims at integrating smart signal control with direct vehicle-to-infrastructure and pedestrian-to-infrastructure communication to enhance safety and mobility. He is a Fellow of the Association for the Advancement of Artificial Intelligence (AAAI), and he has published over 270 technical articles in this general area.

Aaron Steinfeld PhD is Associate Research Professor in the Robotics Institute at Carnegie Mellon University and the Co-Director of the Rehabilitation Engineering Research Center on Accessible Public Transportation. He earned his PhD, MS, and BS in Industrial and Operations Engineering from the University of Michigan (1999, 1994, and 1993, respectively) and completed a postdoctoral position at the University of California, Berkeley (2000). Steinfeld's interest is focused around constrained user interfaces and operator assistance, predominantly in the realms of human-robot interaction, rehabilitation, transportation, and intelligent systems. He is interested in how to enable timely and appropriate interaction when interfaces are restricted through design, tasks, the environment, time pressures, and/or user abilities. Two core areas of interest are: Human-Robot Intent Fusion and Appropriate Robot Behavior. Examples of past work includes human-machine interaction and interfaces for collision warning systems, drowsy driving, zero-visibility snowplow operations, advanced in-vehicle systems, head-up displays, real-time captioning, rehabilitation robotics, military robotics, semi and fully autonomous mobile robots, and software agents.

Edward Steinfeld ArchD, AIA is a registered architect and gerontologist with special interests in universal design, accessibility, and design for the lifespan. He is a SUNY Distinguished Professor of Architecture at the University at Buffalo, where he has been on the faculty since 1978. He received his Bachelor of Architecture degree from Carnegie Mellon University (1968) and a Masters in Architecture and Doctorate in Architecture from the University of Michigan (1969 and 1972). He is the founding Director of the Center for Inclusive Design and Environmental Access (IDeA), a leading center of excellence in universal design. He is the Project Director of the Rehabilitation Engineering Research Center on Universal Design and the Built Environment grant and co-director of the Rehabilitation Engineering Research Center on Accessible Public Transportation grant. He has over 100 publications and three patents. Many of his publications are key references in the fields of accessible and universal design. He was a co-author of *Universal Design: Creating Inclusive Environments*. He serves as a consultant to government and industry, and he has received many awards for his contributions to the field.

Heamchand Subryan MArch/MFA is an expert in introducing responsive information technologies into built environments. At the IDeA Center, his responsibilities include interactive technology development, and graphic, website, exhibit, and interactive design. He has extensive experience with computer-aided design and 3D visualization software such as Rhino, 3D Studio Max, AutoCAD, and rapid prototyping machines including 3D printers, laser cutters, and Computer Numerical Control (CNC). As a researcher, he is interested in interactive technologies for the disability population. His recent work includes developing a commercial prototype product of a "touch model" that helps orient blind and visually impaired users in their environment and developing a prototype smart wayfinding system that uses Near Field Communication (NFC) technology and smartphones to enable the blind community to hear contents of signage in the environment. He is also involved in research on the evaluation of commercial products, information design, and wayfinding in built environments.

Anthony Tomasic PhD is a Senior Systems Scientist in the Language Technologies Institute at Carnegie Mellon University. He received an undergraduate degree with honors in Computer Science from Indiana University and a PhD in Computer Science from Princeton University. In 1999, he participated in a team that was a winner in the French National New Venture competition. He then spent 3 years with various Internet dot bomb start-ups in Silicon Valley. Eventually he moved back into research at Carnegie Mellon University. In 2009, he received an MBA from the Tepper School of Business at Carnegie Mellon University. In 2011, in partnership with three other faculty, he founded Tiramisu Transit, LLC. His research now focuses on applied machine learning, mixed-initiative interfaces, and databases.

Daisy Yoo is PhD candidate in the Information School and a member of the Value Sensitive Design Research Lab at the University of Washington. Her work spans the fields of design, human-computer interaction, and information science. In particular, she is interested in the use of digital technologies to support public dialogue and action in politically contested arenas. The focus of her thesis work is on addressing challenges of designing with emerging, pluralistic publics in the case of legalizing medical aid-in-dying in the U.S. Prior to University of Washington, she received her Master's in Interaction Design from Carnegie Mellon University.

John Zimmerman MDes is Professor in the Human-Computer Interaction Institute at Carnegie Mellon University. His research focuses on inventing new ways for people to interact with intelligent systems, on operationalizing identity theory in the design of new interactive products and services, and on how design can be used as a mode of inquiry for scholarly research. He has expertise in interface and interaction design, user experience design, service design, social computing, mobile computing, and ubiquitous computing. He has published more than 100 papers and 30 patents. He teaches courses in user experience design, design theory, and service innovation. Prior to joining Carnegie Mellon University, he worked at Philips Research on advanced concepts for personalized, interactive television.

Preface

Public transportation plays an important role in creating an accessible society because it is critical for ensuring employment, completing activities of daily living, engaging in citizenship, and participating in social roles and social interaction. Accessible public transportation in the community allows individuals with disabilities, especially those with severe disabilities, to have independent access to work sites, educational programs, health facilities, voting and other citizenship activities, and social and recreational activities. In a mobile, global culture, full social participation hinges on accessibility of transportation systems. This book summarizes the "state of the science" in accessible public transportation at the community level rather than intercity transportation like air, long-distance train, and bus coach service. The focus is on "intra-city" transportation because it is the highest priority for people with disabilities. Many sources of information were used to develop this text including systematic literature reviews, direct participation of many experts in the field at conference events, a national online survey, in-depth usability studies carried out by the authors, and site visits to several major public transportation agencies. This is the first attempt to provide an in depth overview and evaluation of research and practices in this field.

In the U.S. the primary mode of travel is the automobile. Since the Second World War, most metropolitan areas in the U.S. have undergone rapid and significant expansion based on two conceptual models of urban growth, the high-capacity highway and single-use zoning practices that were embraced in the U.S. after the War. Today we know that these two practices led to widespread urban sprawl and a heavy reliance on automobile transportation. The emphasis on automobile transportation also reduced demand for public transportation and led to severe underfunding of transportation providers. People with disabilities who cannot drive and low-income citizens who cannot afford an automobile are dependent on public transportation. But many communities have inadequate service, particularly rural and suburban areas. Service limitations increases reliance on expensive paratransit which is mandated by federal disability rights laws. This creates a significant burden for public transit agencies. Public policy, at least at the planning level, is now shifting toward increasing emphasis on public transportation, as fuel prices rise and ridership increases steadily. The aging of the population will increase demand but it will also increase the burden of paratransit service for providers. Other countries face similar challenges.

Transportation should be conceived as a service, which includes the infrastructure of the system (e.g., rail lines, stops and terminals, etc.), management and operations practices (e.g., service routes and schedules, fare reduction policies, eligibility requirements for paratransit, types of vehicles purchased, driver training, etc.), and information and communications (e.g., trip-planning resources, announcements on vehicles and in terminals, real-time arrival information, rider training, etc.). Public transportation service practices are highly diverse as are the provision of accessibility services such as design and construction guidance, staff training, rider outreach and training, fare reduction policies, fare payment practices, etc. What constitutes an inclusive public transportation system? It is clear that the first criterion is that service is available and provides access to the destinations that people need to reach (e.g., worksites, shopping, health care settings, education settings, recreation settings, etc.). The second criterion is that it is affordable due to the relationship between disability and low income and the inability of many people with disabilities to drive. The third is that the elements of a system must be usable for people who have limitations to mobility, perception, and cognition.

The needs of people with disabilities are not unlike those of other public transportation riders. In a cultural context like the U.S., where public transportation has been underfunded, agencies have focused their priorities on service delivery to commuters. But, at a time when ridership is increasing, a more consumer-oriented approach is needed to ensure satisfaction with service in comparison with automobiles and taxis and, for those who are eligible, paratransit. The concept of universal design provides a framework through which improvements in accessibility can also benefit other passengers. The best example of universal design in transportation is the low-floor bus, which is now the standard for buses throughout the U.S. Despite the general favorable reception toward universal design, the concept has not yet been widely adopted in the transportation industry. Many providers simply do not know about universal design or believe that it will result in additional costs. There has not been any research to document examples of universal design and study the economic costs and benefits. The costs of not adopting universal design approaches also need to be investigated.

Despite the high investment of public funding and significant mandates through regulations, there is little systematic research on the effectiveness of different practices of accessibility service delivery. Moreover, there is a great deal of innovation happening at the local level that is not widely disseminated in the public transportation community, restricting the potential for increasing accessibility. A major problem in service delivery is the coordination of accessibility services and policies across different providers in a region. The available evidence indicates that current regulations for transportation are not sufficient to ensure a fully accessible transportation system because the travel experience extends beyond the scope of the issues covered in regulations. The Travel Chain is a useful framework for studying accessibility and providing accessible services. This concept has an end user focus and recognizes that any element in the travel experience, from origin to destination, can restrict accessibility. Elements of the Chain include trip planning, travel to station, station/stop use, boarding vehicles, using vehicles, leaving vehicles, using the stop or transferring, and travel to destination after leaving the station or stop. A key research and development issue is solving the "last mile" problem, getting to and from transit stops and stations. The greater the difficulty in achieving this goal, the more people will be dependent on paratransit. The pedestrian environment leading to stops and stations has been identified as a key problem in providing a fully accessible Travel Chain. Using the Travel Chain concept as a framework, four key research and development areas in accessible transportation related to technology issues are trip planning, the built environment, vehicle design, and wayfinding technologies.

Primary methods used for providing trip-planning information are printed media, help lines, websites, and smartphone applications. While there has not been a comprehensive survey of how trip-planning information is currently distributed, it is likely that all large transit systems now provide web-based trip planners and travel information on their websites as the primary outlet for rider information. Smartphone apps are increasingly available to access this information but other smartphone apps provide additional information sources. The transition to digital information methods for trip planning is accelerating. This is beneficial for people with disabilities since it offers the flexibility to provide the same information in whatever form an individual needs (e.g., text, speech, tactile, at relatively low cost, and, it provides the opportunity to deliver trip-planning information wherever a rider is located and whenever they need it). But, currently, the degree of accessibility in the resources provided is limited.

The built environment plays a very important role in the accessibility of public transportation systems. The built environment includes: (1) pedestrian paths to stops and stations; (2) local stops; and, (3) stations and terminals. Most accessibility issues related to the built environment are similar to those for buildings and facilities as a whole but there are specific issues in transportation that require knowledge beyond that contained in accessibility codes and standards. The research on this topic is limited. There are some research studies of bus stops and shelters and many manuals that transportation agencies use to plan stops. There are many good ideas and innovative practices, particularly in solving the problem of the horizontal and vertical gap at station platforms. Information needs in transportation terminals have received less research attention.

The key research and development areas in accessible vehicle design are related to boarding and disembarking methods and interior design. Boarding and disembarking vehicles can involve multiple interrelated steps: negotiating a level change from grade or platform to vehicle floor at ingress, fare payment, interior circulation down a primary aisle, finding and taking a seat (or, in the case of wheelchair users, securement in a designated area). There has been some research in this field but primarily in other countries where practices and policies are different. Research by contributors to this book on ramp slope demonstrates that current U.S. policies and guidelines are inadequate. Research on the accessibility of interiors of vehicles is a systems problem in which fare payment systems, entry location, seating arrangement, and fixed vehicle features play an important role. Since transit providers have different goals and operate in different contexts, a single best solution is less than optimal. Research on this topic identified alternatives to current practices that can increase accessibility significantly.

On-demand transportation, provided by taxis, ride sharing services, paratransit, and shuttle buses, are an increasingly important part of public transportation systems. Policy developments, both in the U.S. and abroad, have focused mostly on paratransit and accessible taxi service. However, the advent of ride sharing enabled by crowdsourcing technologies and mobile phones, and the ubiquitous use of shuttle buses to fill gaps in the fixed route transportation systems, is focusing attention on these modes of transportation as a public resource. Further, they offer low-cost alternatives to expensive public transportation services like paratransit.

Robotic technology is on the horizon and could rapidly transform the entire transportation system. In particular, these new advances offer solutions to serving an aging population that will be a major challenge in the near future as the baby boomer population matures. Further, these technologies offer options for developing countries and rural areas where current transportation options are very limited. Advanced technologies present many new issues that need to be addressed to ensure that they will benefit people with disabilities as well as other members of society. These include improving accessibility of software applications, accessibility to smaller vehicles, driver training, and service policies.

Wayfinding in transit systems has generally been aided by signs identifying vehicles and stops, direction signs, and fare schedules. Other important sources of information are help lines and information desks. But, as in trip planning, digital-based localization systems are now providing a much richer level of assistance for all travelers and can be especially beneficial for people with disabilities, if they are designed to be accessible. Digital forms can overcome problems with the traditional forms like general inaccessibility of graphic information and text and the poor reliability of human information sources. Current information technology infrastructure, however, presents limitations to the realization of the full potential of digital information technologies. New technologies are under development that can address these problems.

One of the most important advances in information technology for the future of accessible transportation could be social computing, a research area that investigates how new communication systems can be designed to engender specific behaviors. Recently, this research community has become interested in how people and computing systems can be combined to form socio-technical systems that can do things neither computers nor people can do independently. Another IT advance is the concept of "service design." Service design implements techniques like co-design with users and "eGovernance" to improve service delivery. A new accessible next bus application for smartphones, developed by this research team, incorporates both social computing and service design. Preliminary evaluations demonstrate that such tools have great promise in improving usability of transit systems and provider operations.

Accessible public transportation should be viewed broadly as a service design problem. Conceived in this way, the effectiveness of a system should be measured by the experience of the riders. This new approach takes a consumer perspective in which the goal is to address the riders' needs effectively rather than just improve a system's level of compliance with design standards and rules. There are significant problems evaluating service delivery since experiences can change from place to place, day to day, and even minute to minute. This variability presents a difficult problem in evaluation, but new ideas and technologies offer some promising solutions.

The state of the science in accessible public transportation is rapidly evolving due to economic pressures to improve public transportation services with greater efficiency, advances in both vehicle and information technology and new concepts of service delivery developed by innovative providers. Key challenges for the near future include:

- developing seamless systems for vehicle boarding and disembarking;
- improving the pedestrian environment of stops and stations;
- developing strategies to serve low-density communities and neighborhoods;
- providing real-time information in accessible formats;
- improving service quality in general and accessibility services specifically;
- providing location-based information in accessible formats;
- providing innovative and accepted forms of wheelchair securement.

In addition, there are several challenges that researchers and practitioners need to take into account as the field develops further:

- the aging baby boomer generation will increase the number of people who are dependent on public transportation;
- the obesity epidemic that is increasing service difficulty;
- serving rural areas and suburbs, each of which has their own characteristic problems and needs;
- changes in energy policy that may lead to radically different forms of public transportation more amenable to the use of alternative fuels;
- the shrinking of the digital divide caused by rapid adoption of smartphones by all generations.

Acknowledgments

The editors are indebted to several colleagues who contributed significantly to the production of this book. Heamchand Subryan and Ellen Ayoob contributed to Chapter 4. James Lenker, Clive D'Souza, and Victor Paquet wrote Chapter 6. Zachary B. Rubinstein and Stephen F. Smith wrote Chapter 8. Anthony Tomasic and Yun Huang contributed to Chapter 9. Daisy Yoo and John Zimmerman contributed to Chapter 11. In addition, Heamchand Subryan found, compiled, created, and managed many of the figures and graphics.

The development of this book was partly supported by the Rehabilitation Engineering Research Center on Accessible Public Transportation, a center of excellence grant (#90RE5011-01-00) provided by the National Institute on Disability, Independent Living, and Rehabilitation Research (NIDILRR). NIDILRR is a Center within the Administration for Community Living, Department of Health and Human Services. The contents of this book do not necessarily represent the policy of NIDILRR, ACL, HHS and you should not assume endorsement by the Federal Government.

Photo of a sidewalk with pedestrians and a nearby metro bus.
Source: Douglas Levere

1 The Importance of Public Transportation

Aaron Steinfeld and Edward Steinfeld

Overview

Public transportation plays an important role in creating an accessible society because it is critical for ensuring employment, completing activities of daily living, engaging in citizenship, and participating in social roles and social interaction. Accessible public transportation in the community allows individuals with disabilities, especially those with severe disabilities, to have independent access to work sites, educational programs, health facilities, and social and recreational activities. In a mobile, global culture, full social participation hinges on accessibility of transportation systems.

However, the current state of accessible public transportation is a barrier to social participation and, particularly, employment. Approximately 6.5 percent of the U.S. population is 65 years and older, while more than 20 percent of the entire population has at least one disability (U.S. Census Bureau, 2006). More than half a million people with disabilities cannot leave their homes because of transportation difficulties (BTS, 2003). Even when they are able to leave their home, one-third of people with disabilities have inadequate access to transportation (NOD, 2004). Consequently, four times as many people with disabilities as people with no disabilities lack suitable transportation options to meet daily mobility needs (NCD, 2005).

The consequences of inadequate access to transportation are severe. According to one study, 46 percent of people with disabilities, compared to 23 percent of people without disabilities, reported feeling isolated from their communities (NOD, 2000). Individuals with disabilities were five times more likely to report dissatisfaction with their lives than were their non-disabled counterparts, and a majority of those surveyed said that lack of a full social life was a reason for this dissatisfaction. Persons with disabilities were about half as likely to have heard live music, gone to a movie, or attended a sporting event or concert over a 1-year period (Hendershot, 2003; NOD, 2000). People with disabilities, in both urban and rural areas, frequently cite a lack of local transportation as hindering their ability to find employment. Lack of transportation (29 percent) was only second to a lack of appropriate jobs being available (53 percent), as the most frequently cited reason for being discouraged from looking for work (Loprest & Maag, 2001).

Unfortunately, public transit is facing significant funding challenges throughout the country, leading to significant cutbacks and reductions in service. The challenges of public transportation in general, combined with the critical importance of local transportation for people with disabilities and older adults in terms of social participation and employment, proves that there is a real need for innovation and knowledge that can help transportation providers serve all citizens more effectively and efficiently.

Why Is This Important Now?

There is evidence that the relative advantage of public transportation over private cars is changing, and not just for commuters. The time is at hand to seriously address convenience, comfort, and safety issues so that riding transit will not only be economically attractive but a *greater value*. Changing demographics mean that there is a particular need to address the needs of current older adults and the Boomer generation in addition to younger riders (Foot, 2007; Koffman, Raphael, & Weiner, 2004).

It is important to note that improving access to transportation has economic benefits as well as costs. Investing in public transit has positive financial impacts associated with air quality, productivity, leisure activities, and land use. Accessible public transit can fill the gap when independent driving is no longer safe, leading to potentially large safety and health-related savings. The loss of driving ability is a major trigger of relocation to long-term care settings. If public transportation can help delay the national average for such relocation by just one month, it could save $1.12 B annually (Johnson, Davis, & Bosanquet, 2000).

Although the proportion of older drivers is increasing, nearly seven million persons 65 and older do not drive, either due to health or financial deterrents. Older adults with lower incomes are particularly effected by barriers in public transportation systems because they rely on public transportation more regularly (Houser, 2005). One in seven non-drivers age 75 and older currently uses public transportation as their primary mode of transportation (Ritter, Straight, & Evans, 2002).

The Right Information at the Right Time

Many of the problems associated with providing effective and efficient public transportation stem from a lack of relevant and timely information, including schedules, route maps, route instructions, information on the real-time status of vehicles, and information about temporary problems along the intended route, which alerts riders to the need for finding another route. Due to the complexity and large geographic reach of public transit, without good quality information provided when they need it, riders often are unable to obtain effective situational awareness. This leads to an unhappy consumer experience, fear, confusion, anger, and reduced interest in continuing to use transit. This problem is well known to transit agencies. Transit agencies themselves also need better situational awareness about the status of their systems in order to adjust quickly and seamlessly to service challenges.

Many transit agencies are now actively pursuing novel methods for improving situational awareness through adoption of new information technology. For example, increasing numbers of transit agencies are acquiring automatic vehicle location (AVL) systems to provide better service. These systems allow the providers to track vehicles in real time and make adjustments rapidly, but, they also establish a potential to deliver real-time arrival estimates. Providing real-time estimates alone can increase ridership on some routes as much as 40 percent (Casey, 2003) and close to two percent system-wide (Brakewood, Mcfarlane, & Watkins, 2015). Such data is particularly important to people with disabilities. This group is generally more vulnerable to exposure in severe climates, often have medical needs that require timely attention, and have heightened concern about security risks while waiting at stops. The value of situational awareness is demonstrated by research findings for riders without disabilities. For example, use of a real-time arrival information system improves perceptions of security (Ferris, Watkins, & Borning, 2010a). Unfortunately, real-time arrival systems are expensive and often beyond the reach of cash-strapped agencies. Another important information need for riders is knowledge about system component maintenance problems. Few agencies provide riders such information and it is generally limited to elevator and escalator status. Another type of information that can benefit riders, even before they leave their origin for a stop, is information about whether the next vehicle is too full to board or find a seat. "Fullness" data is extremely rare in transit systems.

Applying a Need to Knowledge Model

Lane and Flagg (2010) outline a process that links research to development activities. They argue that research is most effective if it is tied to needs for knowledge that arise during the development process. The delivery of public transportation service is a development activity supported by many sources of information. Important decisions in the early stages of service planning and design cannot be undone, often for decades. Thus, an effective program on accessible public transportation should pay close attention to the related information needs of the industry.

Significant information needs about accessibility exist in vehicle and station design, service planning, and operator training. Important details about best practices, the needs of users with specific disabilities, the interaction of environmental factors and service delivery, and other issues are often hard to find or still yet to be studied by the research community. In general, the field has limited channels for sharing information, relies primarily on practical experience at the local level, and on anecdotal or case data rather than systematic approaches to information as part of the development process. Thus, the need for relevant research and development that is closely tied to practice is critical.

Currently, there are some well-developed strands of research within the field of accessible transportation. But, there are also many sources of information that can be applied to accessible transportation problems that are not currently being utilized. In particular, knowledge bases and ideas from the fields of human factors, rehabilitation science, and interactive design are not fully utilized. One example of the gap is the design of accessible information kiosks. For years, information has been available on ways to make such information accessible to people with a broad range of disabilities (see, for example, Subryan, Landau, & Steinfeld, 2012; Vanderheiden, 1997; Vanderheiden, Law, & Kelso, 1999).

The gap in application of available knowledge is an important issue that has not been recognized in the literature. Just as research should be tied to needs for knowledge, there is also a need to identify, organize, and disseminate existing knowledge in a way that providers can make effective use of it. This process should start with an understanding of how development takes place in the industry, what needs for knowledge exist and are currently unfilled, and in what form knowledge should be provided to be of most use to the industry. We hope that this book will help bridge the gap between research and practice, thus fostering better understanding of industry needs and raising awareness of systematic research that is ready for practical applications.

Photo of a suburban bus loop that has tourists gathering near a large bus shelter.
Source: IDeΛ Center

2   The Culture of Accessible Transportation

Edward Steinfeld

Overview

Social, economic, and political developments and trends combined with related practices create the culture, in the terminology of anthropology, within which accessibility to public transportation is achieved. In this chapter, we will examine this culture—the larger social context in which accessible transportation is embedded, and, the policies and operations that are used to implement it. This will help to understand the current status of accessible transportation, which intervention strategies are likely to be most effective, and how best to implement them.

Important developments and trends include national transportation policy, housing and urban development patterns, and demographics. Important practices include the policies governing the operations of public transportation agencies, and the business of purchasing, designing, and maintaining rolling stock and infrastructure.

Urban and Regional Development

Since the Second World War, most metropolitan areas in the U.S. have undergone rapid and significant expansion based on two new conceptual models of urban growth, the high-capacity highway and single-use zoning practices that were embraced in the U.S. after the War.

In the high-capacity highway model, multi-lane highways were constructed in radial patterns from the center of cities to the outlying suburban areas and around the urban core in circumferential patterns, or "beltways." To maximize capacity through high speed limits, the highways were constructed with limited access. Unlike the traditional multi-lane boulevard, where all intersections were on grade, high-capacity highways only have access at "intersections," which are built so that roads pass over each other either above or below grade. The growth of this highway network was financed in two ways, use of tax money to build "freeways" or private financing, through bond issues of government authorities to build toll roads (Weingroff, 2011). Later, the federal gas tax was used to build the interstate highway system, some of which is overlaid onto older toll roads and freeways. The impact of these practices, particularly, the funding of "free" highways through taxes, made commuting by automobile to new suburbs very inexpensive when gas prices were in the 25 cent/gallon range and parking costs were either very low or provided free by employers, civic authorities, and retailers. Thus, middle-class families who lived in city centers and new immigrants from rural areas could afford to move out of the older city centers and commute to their jobs from suburban locations. Businesses soon followed the exodus from the city centers because they also could take advantage of the inexpensive and efficient highway transportation system (Baum-Snow, 2007).

In the single-use zoning model, municipalities at the edge of central cities zoned their land for future development in a way that limited what could be built in each part of their jurisdiction to one predominant use (e.g., single-family residential, multi-family residential, commercial, light industry, heavy industry, civic, etc.). The goal of these zoning practices was ostensibly to avoid the "incompatible" land uses often found in older cities where a factory could be located right next to a residential neighborhood. Typically, the zoning pattern followed the pattern of highway construction. The overwhelming pattern of zoning was single-family detached housing strung out along residential streets off "collector" streets leading to commercial strip centers. Commercial and industrial land uses were located near intersections. Multi-housing zones were limited to the suburbs and often located in undesirable land next to highways between intersections or adjacent to commercial zones.

In the first ring post-Second World War suburbs, relatively modest houses and lot sizes and grid patterns of streets were built to accommodate returning veterans and their new families. The subdivision plans of second ring suburbs often abandoned the urban grid pattern to reduce traffic on residential streets. By that time, the new houses had increased in size with commensurate lot sizes. In the development of third ring suburbs, starting in the 1980s, cul-de-sacs were used extensively, further reducing traffic in residential areas. Subdivisions in the latest wave of development have limited entries, are often gated, and are separated from the neighboring subdivisions. Even today, many suburban streets are constructed without sidewalks. These development patterns concentrate traffic on collector streets, reduce pedestrian activity to almost nothing, and increase the length of any trip, compared to the urban grid system.

Today we know that these two practices led to widespread urban sprawl and a heavy reliance on automobile transportation, for almost every resident of the suburbs, including children. In fact, the lack of sidewalks in many suburbs makes it dangerous for children to walk to school, recreation sites, and even to visit friends in nearby subdivisions.

Thus, at an early age, citizens become acculturated to automobile transportation as the norm for getting around their communities. Presently, about 70 percent of all Americans live in single-family homes (U.S. Census, 2001). Sprawl had the side effects of highway congestion, increased pollution, and sedentary life styles, but living in sprawl has been the normal experience for generations of adults, most of whom have literally no experience with public transportation.

There are many other factors that contributed to sprawl such as government insurance for mortgages, tax, and monetary policies that kept the price of housing affordable; increasingly less stringent mortgage qualification practices that fueled housing booms; white flight from inner cities; the availability of free parking in suburban commercial districts; and, big box merchandizing. However, the two practices outlined above were the key factors that led to the current urban and regional settlement pattern. It should also be noted that rural areas did not benefit from these practices except if they became sites for expanded urban development. During the 1970s, it became apparent that public transportation could no longer be ignored, especially to relieve congestion and air pollution. To fund the development of new public transportation infrastructure and maintenance of existing systems in older cities, the federal government initiated a policy of diverting some of the money from the federal highway fuel tax to public transportation.

With urban sprawl and an automobile-oriented culture, ridership on most public transportation systems decreased to the point where they could no longer exist without public subsidy. Fares were increased steadily to the point where any further increases result in loss of ridership, either because low-income riders cannot afford the cost or, the fares make mode shift to automobiles attractive to suburban and more affluent riders. Except for a few older cities like New York, Chicago, Philadelphia, Boston, Washington, and San Francisco, public transportation does not play a major role in the mobility of Americans. Public transportation in these older cities have benefitted from an infrastructure that makes driving both difficult and expensive, an historic familiarity with public transportation use, and an economy that contributed to the prominence of their inner cities as a work location (e.g., financial, education, health care, legal, and media industries). To reduce costs, the operations of public transportation, even in these older cities, have been focused on serving the needs of commuters. This policy determines the routes that are operated, the frequency of vehicles on the routes, the times that service is available, the station locations, and other service characteristics.

There have been many new innovations in public transportation and there are also some promising trends that lead to a better future for this sector of the transportation industry. For example, outer suburbs are now declining as the preferred location of residence due to rising gasoline prices (ULI, 2009). Trends in many regions are shifting from suburban to urban living. Outer ring suburbs are the areas that are suffering most from foreclosures and depressions of the housing market. In fact, the most successful form of suburban development over the last two decades, based on housing appreciation and sales, has been "Traditional Neighborhood Development" or TND (Eppli & Tu, 1999). This form of urban development utilizes a mixed-use approach to land development, emphasizes walkability in planning and better street design, and incorporates transit-oriented development.

Studies of urban development have found that cities that attract a young "creative class" (Florida, 2002) obtain an edge in economic competition (Dwyer & Beavers, 2011). Younger people, often suburbanites by birth, are eschewing suburban living and moving to inner city locations. The difficulty in obtaining mortgages is increasing demand for rentals, which are more plentiful and more varied in higher density urban areas. Empty nesters, the Boomers whose children have left home, show an increasing interest in living in high-density urban areas where they can take advantage of walkable neighborhoods and cultural and entertainment opportunities not available in the suburbs. Trends in driving participation are actually on the decline as more and more teenagers are opting not to obtain a driver's license at the minimum legal age (UMTRI, 2011). Transit ridership has steadily increased over the last six consecutive quarters (APTA, 2012). There is also a growth in active transportation (e.g., walking and bicycling). Many cities are working on how they can improve safety and convenience for bicyclists. The growth of bike sharing programs is particularly noteworthy since increases in bicycle ridership are related to increases in transit ridership. Many public transit systems are now developing their own bike sharing programs because of this relationship.

At the same time as this renewed interest in urban living develops, the suburbs, particularly the first and second ring, are aging. There is a strong relationship between driving fatalities and age when controlled for miles driven (Tay, 2006). Fatal incidents are as high between 65 and 80 as they are in teenage years, although for different reasons. Above age 80, the rates are much higher than for teenagers. Even though older people tend to continue driving as long as they possibly can, there comes a time when they cannot drive anymore. Yet, surveys consistently demonstrate that they want to "age in place" rather than move out of their homes and single-family neighborhoods to locations where driving is not necessary. The effort to control health care costs, particularly by federal and state governments, will lead to increased emphasis on home care and ambulatory care. Thus, as the population ages, these trends will increase demand for different public transportation options, especially if the price of gasoline continues to increase since higher driving costs impact people on fixed incomes more than others.

Problems

- Urban and regional development patterns favor automobile transportation and make public transportation a last resort option for most of the U.S. population, with the exception of some older cities.
- Development patterns are shifting toward higher density, mixed-use development, but a large percentage of the U.S. population will continue to reside in conventional suburbs.
- Several factors, including highway congestion, air pollution, gasoline prices, housing trends, and demographic trends are converging to increase interest and demand for public transportation.

Recommendations

- Utilize the large body of existing research on transportation-related urban and regional development issues and transportation trends.
- Acknowledge and address the disparities in public transportation availability between rural settings compared to urban settings and their impact on people with disabilities and elders.
- Explore accessible transportation needs in suburban areas, particularly with respect to aging in place.

Directions for Future Work

- Trends in urban and regional development that support increased attention to accessible transportation need to be identified and communicated to the transportation industry.
- Differences in needs between urban, suburban, and rural settings need to be addressed.
- Innovative solutions are needed for rural and suburban settings where low density of population distribution presents both economic and operational challenges for public transportation.

Public Transportation Agency Practices

Public transportation is much more diverse than it seems at first glance. Our conventional view of public transportation is an urban transit system, perhaps offering bus, subway, or light rail service, or some combination. Such service is usually provided by a regional quasi-government agency with major investments in stations, tracks, and rolling stock. But, public transportation also includes local transportation providers, owned and operated by government entities. These public agencies provide service only within the boundaries of their towns or counties that operate them. Private companies also play a role, often serving niche markets like small towns that are not served by regional systems or local government. Another type of public transportation provides specialized services targeted at specific markets. These include commuter bus, rail, or ferry service designed primarily for commuters. There are also even more specialized operators serving limited destinations and populations (e.g., airport shuttles, university campus transportation, senior citizen transportation). Larger systems may have their own maintenance operations, repair shops, and even a real estate development operation. Historically, older transportation companies have extensive landholdings, sometimes including commercial facilities they have developed and through which they earn income. The smallest operations may not even own their own vehicles, relying on vendors to furnish and maintain their vehicles.

During the course of the Rehabilitation Engineering Research Center on Accessible Public Transportation (RERC-APT) grant from the National Institute on Disability, Independent Living, and Rehabilitation Research (NIDILRR), co-directed by the author, the research team visited several large regional transportation agencies to learn more about accessible public transportation practices directly from the operators. These visits included riding the systems, guided tours of the systems by agency staff, and meetings with administrators. The focus was fixed route services although we did discuss the relationship of such service with paratransit operations and other local and regional providers. We visited the Niagara Frontier Transportation Authority serving the Buffalo-Niagara Falls metropolitan region (NFTA), the Port Authority of Allegheny County in Pittsburgh (PAAC), the Massachusetts Bay Transportation Authority serving the Boston metropolitan area (MBTA), the Metropolitan Atlanta Regional Transportation Authority (MARTA), the Los Angeles County Metropolitan Transportation Authority (LA Metro), the Washington Metropolitan Area Transportation Authority (DC Metro) and the Southeast Pennsylvania Transportation Authority, i.e., Philadelphia (SEPTA).

While traveling, we also took the opportunity to be riders of many other public transportation services including the Muni, BART, and Caltrain in the Bay Area, the MTA, PATH (serving Northern New Jersey) and the Long Island Railroad in the New York City Tri-State area. Internationally we traveled on public transportation extensively in Europe and Asia, Australia and New Zealand, and to a lesser extent in Latin America. We have also attended many conferences and followed the literature on general public transportation issues.

These visits and experiences helped us understand the current practices in accessible public transportation in the U.S., including the ability to contrast them with practices in other countries, although we are not yet knowledgeable about policy, funding, and operations in those countries. These practices are summarized below.

In all public agencies in the U.S., there must be an individual designated as the ADA/Section 504 Coordinator. That individual is responsible for compliance with the ADA and Section 504. But, each agency has the freedom to decide how best to staff and manage accessibility related services. We will use the term "accessibility services" in reference to the full set of programs that implement accessibility requirements and additional services. This can include design review of construction projects, training of drivers and station staff, maintenance of equipment related to accessibility, fare reduction programs, public education on accessibility including outreach and training of riders, etc.

We observed major differences in the organization of accessibility services from agency to agency. The most dramatic difference was the contrast between the LA Metro and the DC Metro. LA Metro, one of the largest systems in the country with great diversity in types of service, has implemented a de-centralized approach. One person from their headquarters staff, the ADA Coordinator, is responsible for policy and program development. But, day-to-day implementation is accomplished by staff distributed through every department of the agency with a related mission (e.g., maintenance, capital projects, customer service, etc.). The Coordinator provides leadership, consultation, and oversight to the departments. The DC Metro, a much smaller system limited to heavy rail and bus services, has a separate department devoted to accessibility services with 20 plus staff. This centralized approach puts responsibility for all accessibility services in the hands of one team. Other systems we visited fell somewhere in the middle between the extremes of complete centralization and complete de-centralization. This finding raises many research questions about organizational effectiveness. Are the differences across agencies related to the differences in size and complexity of the systems, to historical reasons, management style, or lessons learned through practice? How should accessibility services be organized to improve outcomes? What experience and education are needed to provide different services? What in-service training does line staff need? We have not found any research on these issues.

Local, regional, and federal agencies have many business practices that directly affect accessibility of systems, both positively and negatively. Here are some examples from this research:

- The MBTA has a real estate development program that helps to fund accessibility improvements in stations. In downtown stations, they are partnering with private developers to provide additional elevators to older underground subway stations. When new buildings are constructed above stations, the agency negotiates with developers to include elevators that lead from building lobbies to the station entry lobbies. In some cases, a new street-level station entry may be incorporated into the new building separated from the building lobby. In suburban stations, particularly on commuter rail lines, the MBTA sells air rights on their property to developers who want to build office buildings, housing, and other facilities in these desirable locations. The agency negotiates to ensure that this new development increases accessibility to the stations through the construction of overpasses, underpasses, parking, elevators, and other improvements related to station access.
- SEPTA has some suburban commuter rail stations that are actually owned by Amtrak (the national passenger rail system) and share tracks not only with Amtrak but also with freight operators. It has been difficult for the agency to provide accessibility to these stations because Amtrak has not upgraded their properties. There are two reasons for their inaction, lack of funding and the tolerances needed for freight trains to clear passenger-loading platforms. There is a concern that renovating the station platforms to meet the accessibility requirements for passenger rail may introduce impediments to freight traffic.
- "Franchise-bid" programs are used to fund, construct, and maintain bus shelters in major cities. Advertisers bid for projects in which they agree to pay fees and revenues

Figure 2.1 Bus shelter in New York City awarded through a franchise-bid program.
Source: Author

to the cities or transportation authorities in return for advertising rights on shelters and other related street furniture. The advertising agencies pay the full construction and maintenance costs for the shelters. Some cities have adopted this practice and others continue to construct and operate their own shelters, keeping the revenue from the advertising to fund operations. Franchise-bid programs can improve comfort, safety, and accessibility of bus stops but they can also result in disparities between the provision of shelters in low-income and low-traffic areas compared to high-income and high-traffic areas if the bid winners are allowed to maximize advertising revenue without restrictions and oversight (Figure 2.1).

• LA Metro operates service to major airports. However, the Port Authority of LA County, a separate agency, also operates express buses between LAX Airport and key suburban locations called the Airport Flyer, using "over the road" buses equipped with lifts for wheelchair access. Since the Port Authority also operates parking at the airport, which is in high demand, and wants to reduce congestion in the airport approaches, they constructed large modern terminals at their suburban locations, including parking garages and taxi stands to encourage airline customers and employees to avoid driving to and parking at the airport. However, these terminals are not always located where they provide convenient inter-modal access with the LA Metro services. For example, one terminal we visited was two miles away from a major express bus route station. However, off-peak, regular bus service to the station was very infrequent, requiring an expensive taxi ride (not accessible) to the station to avoid an hour-long wait (Figure 2.2).

Figure 2.2 Flyaway suburban bus terminal with nearby parking facility (a) main entry to the terminal, (b) large parking structure on site.

Source: Author

- In Montgomery County, MD, the local transportation agency has a program of outreach to private landowners of strip shopping centers and malls. Through this outreach program they identify locations and projects that can help improve pedestrian access from bus stops to commercial properties, including accessible features like curb ramps and safety and security features like lighting.
- Most agencies have programs to assess riders for paratransit eligibility and to train riders who are unfamiliar with use of public transportation. Training can reduce the burden on agency staff in the field, inform riders of their right to reduced fares, and improve the skills of riders so they can use fixed route services rather than paratransit. SEPTA has developed a training center that is equipped with elements of the transportation system that are difficult to negotiate by people with disabilities like the platform gap, bus entries, fare boxes, securement systems, and bus boarding ramp. Passengers with disabilities can be brought to the center for personalized training and assessment. One goal of this program is to encourage individuals qualified for paratransit to ride fixed route service. It can also be used to train new bus drivers and other line staff (Figure 2.3).

Figure 2.3 SEPTA training facility for riders (a) simulated bus, (b) simulated subway station platform.

Source: Aaron Steinfeld

The examples above are just a sample of the diverse practices we discovered. They indicate that there is a need for research that would identify, document, and disseminate good practices. Furthermore, they indicate that systematic research on practices could identify many problems, like the Amtrak example, that need research, policy, design, and planning attention. Presentations at the Transportation Research Board (TRB) and the International Conference on Mobility and Transport for Elderly and Disabled Persons (TRANSED) conferences include reporting on successful practices but the agency personnel that present these examples are usually not writers or researchers; the presentations are not all recorded and archived for later reference; and, there is no external verification of claims. Few agency personnel attend these conferences, particularly TRANSED, which is usually outside North America. Moreover, identification and documentation of problems such as the Amtrak station issue noted above could demonstrate the need for policy changes or even technology solutions like how to reduce the platform tolerance needed for freight trains running on passenger lines.

Problems
- The effectiveness of different accessibility service approaches is not studied systematically, including the factors of budgetary resources, organizational scale, and experience and education of staff.
- There is no research available on the impact of complexity, in the array of transportation providers in one location, on understanding and using accessibility services.
- Practices developed at the grass roots are not systematically studied and information on innovative programs and best practices is not widely disseminated.

Recommendations
- Explore how service delivery practices are related to accessible transportation.

Directions for Future Work
- Develop a culture of evidence-based practice among accessibility service providers.
- Improve communications between transportation providers on accessibility services.
- Develop methods for benchmarking effectiveness of services and identification of best practices.

Transportation Policy and Regulations

Metropolitan transportation agencies receive extensive funding from their state departments of transportation and the federal government. During the last four years, they have received stimulus funding for capital improvements and vehicle purchases. Since they receive public funding, they must comply with the federal regulations for accessibility, promulgated by the U.S. Department of Transportation (DOT) and the U.S. Access Board (U.S. Access Board, 2009). The ADA vehicle standards, issued by the DOT, address design issues related to vehicles, primarily boarding and disembarking features, fare payment, in-vehicle communications, and interior passenger compartment design, although they also address toilet rooms and other amenities in vehicles like trains and airplanes. For stations and stops, the ADA standards for transportation facilities developed by the U.S. Access Board apply through reference in DOT regulations.

State and local government may also introduce additional regulatory constraints through state accessibility laws or through related regulations. Examples are requirements for accessible parking spaces and curb ramps and regulations for installing structures like bus shelters and bike racks in public rights-of-way.

Currently, the DOT and Access Board develop regulations that are based on input from stakeholders, including consumer advocates, equipment manufacturers, and operators, without verification from empirical research, unless the researchers actually participate in the public input phase of regulatory activity. The two agencies do not always fund research prior to writing proposed regulations to address specific knowledge gaps. Until recently, research devised to develop recommendations for accessibility to buildings in the 1970s was applied in transportation settings. This research never obtained data on many important issues for transportation. One example is the weight of wheelchair users. The increase in obesity in the population has resulted in the need for transit agencies to accommodate many more wheelchair users who may exceed the capacity of lifts and ramps, even if the equipment meets the weight capacity mandated by regulations. Now that the problem has been identified, there is a scramble for research data that could inform regulations and product design on this topic. The RERC-APT research demonstrates that fold out ramps can present some dangerous conditions for visually impaired transit riders but the regulations governing the design of these ramps focus primarily on accommodation of wheeled mobility users. A requirement for an internal public address system has been applied to vehicles in excess of 22 ft in length that are used in fixed route service with multiple stops, but there are no regulations that address the quality of sound, the acoustic environment of the vehicle or the clarity of speech used in the announcements.

Our current research program on boarding and disembarking offers a good example of how research could be conducted in service of regulatory activity. The research team contacted the U.S. Access Board to find out their information needs related to regulatory issues. During our proposal development, we identified two high-priority issues for research that could contribute immediately to improving the state of the art in accessible public transportation, the slope of boarding ramps and space clearances in the front of low-floor vehicles. Independently, after we received the grant, the Board issued a Notice of Proposed Rule Making and currently is holding hearings on changes in the regulations. Our research findings are now available to provide input into this regulatory activity. We expect that our findings will also lead to development of new products to address the accessibility limitations of existing equipment.

Findings from existing research have raised questions about the adequacy of some fundamental practices in the field. One example is the current U.S. practice of securing passengers facing forward with tie-downs. In other countries, protected compartments with rear-facing positions are used for securing wheeled mobility users. No tie-downs are provided because the rear-facing position with a padded wall behind the passenger allows containment of the device during a crash. In addition, the driver does not have to help the passenger get into position. Despite the fact that rear-facing compartments do not address the problem of lateral movement of occupants, this approach to securing passengers deserves consideration in the U.S. Research has discovered that most wheelchairs are not crashworthy when tied down facing forward (Fitzgerald, Songer, Rotko, & Karg, 2007; Frost & Bertocci, 2007). Standards for construction of crashworthy wheelchairs have been developed (i.e., ANSI/RESNA WC 19) but regulatory action has not yet mandated application of those standards since they would require passengers to purchase wheelchairs designed primarily for transportation rather than for ease of use and comfort.

Current regulations require drivers to ask whether a passenger wants to be secured and policies vary by transit agency whether passengers are permitted to ride without securement if they refuse (Frost, van Roosmalen, Bertocci, & Cross, 2012). Moreover, faced with diverse equipment that is not easy to secure using current systems and pressure to stick to schedules, drivers often do not adequately secure passengers. We observed a wheelchair user sitting unsecured with his feet propped up on a wheel well and another who was inadequately secured topple over during a sharp turn. Wheelchair users report that sometimes drivers simply drape the tie-downs over their chairs. Difficulty with securement systems is legendary in the industry. One ex-driver we interviewed obtained a chronic impairment when a morbidly obese power chair user rolled over his foot and he reported that a colleague obtained a chronic back injury helping another up a ramp. Thorough research on actual practices in the field would be useful to understand the limitations of current regulatory practices, equipment designs, and training. Such research should include attention to both rider issues and operator issues to understand all sides of the problems. The securement issue points out that research focused on one issue like safety, in isolation from workplace ergonomics and affordability, may not lead to realistic solutions.

Problems

- Regulations for services and design do not have an adequate evidence base leading to requirements that may create problems while they try to address others.
- Knowledge of innovative approaches is not disseminated adequately throughout the industry.

Recommendations

- Explore ways of compiling existing research evidence and making it easier to access.

Directions for Future Work

- Assembling and evaluating the current knowledge base supporting accessible design and offering services to make that knowledge available to policy developers prior to the development of proposed regulations.

Design and Manufacturing of Rolling Stock

Vehicles are very expensive. For example, an urban bus can cost $500,000–$600,000. Thus agencies want to keep them in operation as long as possible. Some agencies we visited had buses still in operation that were over 30 years old, about 10 years beyond the rated lifespan. Rail cars tend to be older than buses. Accessible features on vehicles therefore have a long-term impact.

We have not studied the rail industry but we know that bus manufacturers design their vehicles to meet the federal regulations and the specifications of their clients (e.g., transportation agencies and companies). Many key parts of each vehicle are produced by equipment manufacturers like seating and lift companies. Thus, transit vehicles are designed and manufactured like computers. They are a basic framework of chassis and an enclosure to which many different parts produced by others are fastened. Many of these parts have to work together so the problem of interoperability is critical for efficient production.

From discussions with equipment and bus manufacturers, we discovered that the design of buses is not very sophisticated by contemporary design standards. For example,

3-D human modeling and full-scale simulation and testing of new features are used extensively in the automobile industry to evaluate ergonomic fit during the design process. But, in the bus industry, human modeling and simulation are not yet routine. Perhaps this is because the parts are basically off-the-shelf items. Manufacturers select from what is available rather than design the parts like seating or lifts as part of the vehicle and then have them made by suppliers to spec. Bus manufacturers and their suppliers learn from experience and feedback from their customers about problems and correct them in newer models. Major changes are difficult to introduce using this design process because of the interoperability problem. Making a radical change on one part can create major problems. New parts have to fit into the system. Thus, for example, widening a door may be restrained by the location of structural members in the vehicle frame.

Transportation companies evaluate the products available from the manufacturers in three ways: (1) they attend the American Public Transit Association's annual showcase event where current bus models are made available by manufacturers; (2) bus manufacturers bring buses to transportation agencies and companies to demonstrate their features; and, (3) agency and company staff visit the manufacturers to evaluate buses.

The bus manufacturing business is a difficult one. The capital investment in building buses is quite high; the reliability demands are stringent and the competition for sales is very tough. Thus there are very few bus manufacturers, even on a global scale. Currently, there is only one truly U.S. manufacturer. Although other manufacturers assemble buses in the U.S. because there are government requirements for a percentage of American content, they are actually based elsewhere. This makes it more difficult to integrate government-sponsored research related to the American market with design and production. Moreover, since more buses are sold overseas than in the U.S., the needs of our market will not receive the same priorities as other markets. For example, mode shift from cars to public transportation is not as important for European, Latin American, and Asian markets because automobile commuting in those countries is not yet the norm. Public transportation providers therefore do not have as much of an incentive to increase comfort and they put more of a priority on bus capacity and affordability.

Problems

- It is difficult to make radical changes to improve accessibility of vehicles.
- The transportation agencies, companies, and federal regulations drive the designs.
- Knowledge about business practices in the vehicle manufacturing industry resides in the manufacturers and the purchasing departments of transportation companies.
- Manufacturers and other stakeholders do not have a method to simulate use of products by people with disabilities during the design process.

Recommendations

- Engage in research to improve the design and evaluation process.

Directions for Future Work

- Develop simulation techniques that can be used both by manufacturers during design and by providers during implementation.

Tools for Practice

There are three tools for practice that, if applied on a widespread basis, could advance current practices in the industry.

The Transect is a tool that can be used to identify and correct urban and rural transit inequalities, classify best practices, guide planning, and inform regulatory activity. Developed by proponents of TND, the Transect is a concept that classifies neighborhoods into six zones by their mix of land uses, density, and general character ranging from undeveloped rural areas to high-density central business districts. The concept is based on the premise that each level of development density requires a different approach to land development and transportation. Its developers propose the Transect as an alternative to single-use zoning that is applied in similar ways throughout an urban realm. In each Transect zone different strategies are used for housing type, street design, transportation service, and other neighborhood elements, in keeping with the density and character of the zone. So, for example, design of streets in the urban core (T6), suburb (T3), and rural (T2) zones should reflect their unique needs. The regulatory environment of public transportation (especially street and roadway design) tends to treat all zones the same, precluding the use of different strategies. But, the service needs, resources, and infrastructure vary significantly from location to location. For example, should we expect that all stops in rural areas be equipped with paved loading platforms? And, if not, then perhaps the design of vehicles needs to reflect those differences also. The Transect can be used to develop appropriate regulations and classify best practices for accessible transportation that address the real differences in physical context, resources, and demand in each zone (Figure 2.4).

The Travel Chain is a concept that can be used to systematically study the effectiveness of accessible transportation practices. It was developed by transportation researchers to communicate the importance of each aspect of transportation systems from the perspective of the individual traveler (Iwarsson, Jensen, & Stahl, 2000). It organizes elements of the transportation environment into sequential activities, starting with the information one needs to plan a trip to arrival at the destination. Like a literal chain, if any element of the Travel Chain fails to support the traveler, then the success of the entire trip is jeopardized. Although a more detailed breakdown is possible, links in the Travel Chain include: trip planning, travel to station, station/stop use, boarding vehicles, using vehicles, leaving vehicles, using the stop or transferring, and travel to destination after leaving the station or stop. If one link is not accessible, then access to a subsequent link is unattainable and the trip cannot be completed (Figure 2.5).

The "Travel Chain" highlights the importance of each element that contributes to the success of a trip. For example, in the early years of accessible transportation, the focus was on vehicle design. Later, the designs of stations and stops were recognized as critical

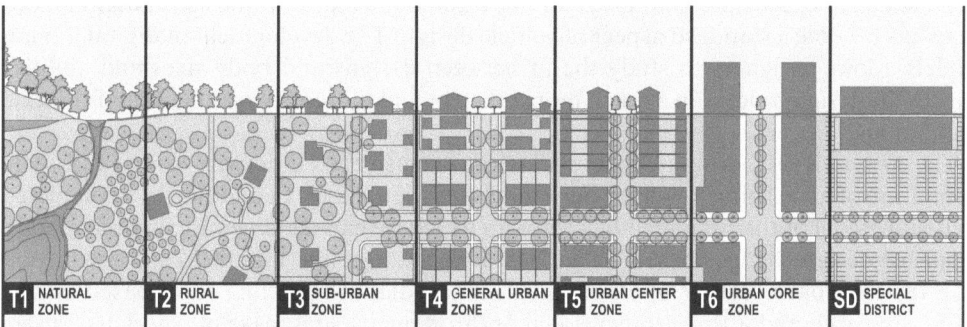

Figure 2.4 Plan view of Transect diagram based on the SmartCode.
Image courtesy of www.transect.org/rural_img.html

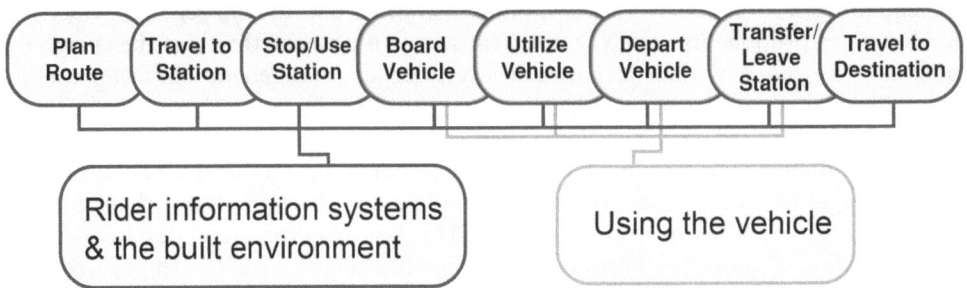

Figure 2.5 The Travel Chain.
Source: IDeA Center

elements. We now understand the importance of rider information systems and pedestrian access to stops and stations for the success of each trip. The RERC's Guided Tours project was developed based on the Travel Chain concept. The results of the project confirmed the importance of each Travel Chain element for all groups of users (Steinfeld, Grimble, & White, 2011). A few examples demonstrate the value of the Travel Chain approach:

Research in the field tends to focus on the information needs of people with vision impairments. But, our findings show that, when physical accessibility issues are reasonably addressed, deficiencies in the information environment become obvious as a major barrier to usability of transit. This includes information for planning the route, information in stations, and information during transit. Moreover, our research indicates that the clarity of the built environment is also an important aspect of rider information. Information systems have to provide information that is clear and understandable. They have to accommodate a wide range of individual abilities as well as a wide range of information needs. Although there are still major research issues to address in vehicle, stop and station design, the information environment is the new horizon. It is especially relevant to accommodate the needs of people with cognitive impairments, mental health conditions, and learning disabilities, groups that transit agencies report are heavy users of expensive paratransit.

Simulation methods can be used to test vehicles in the design stage. The use of full-scale models for testing designs of new vehicles is used extensively in the automotive industry and, to a lesser extent, in transportation vehicle design (see, for example, Daamen, de Boer, & de Kloe, 2008; Petzall, 1993). In the automobile industry, the use of digital modeling has become a standard aspect of vehicle design. The development of digital human models allows designers to study the fit between designs and body sizes and abilities during the design process in a much more efficient and less costly manner than full-scale simulations.

The current RERC program incorporates an extensive R&D activity using a full-scale simulation of an existing low-floor bus model (see Figure 2.6). In addition, we completed a virtual model of the simulated bus through which we can demonstrate, in great detail, proposed design changes based on our research findings (see Figure 2.7). We will soon take this approach further by developing human digital modeling tools based on the RERC research and a three dimensional anthropometric database of wheelchair users we developed in a previous study. These activities are demonstrating the power of simulation to our partners and to others in the industry.

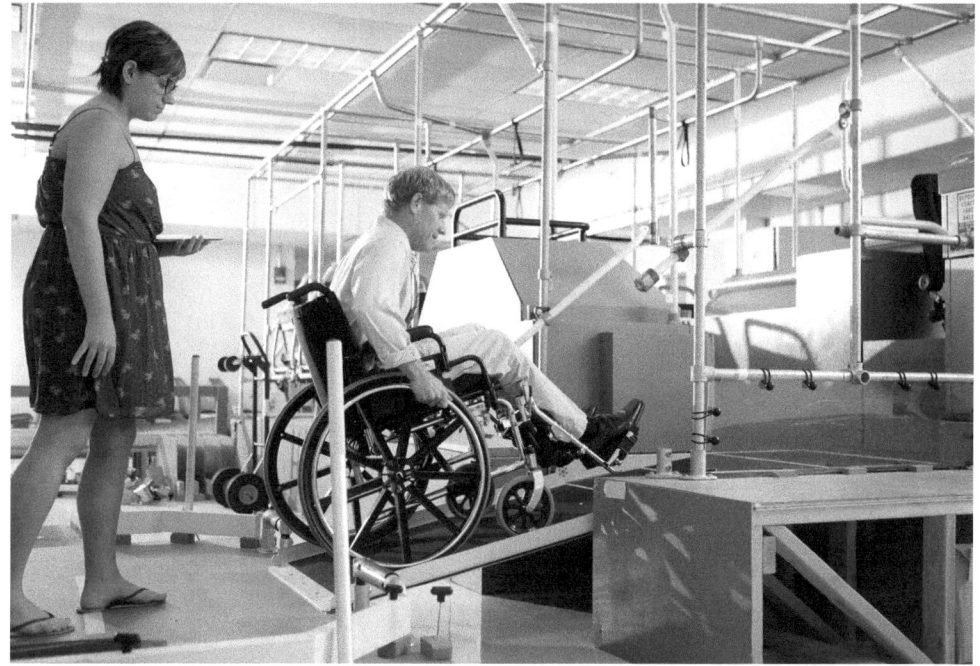

Figure 2.6 Full-scale simulation of existing low-floor bus.
Source: IDeA Center

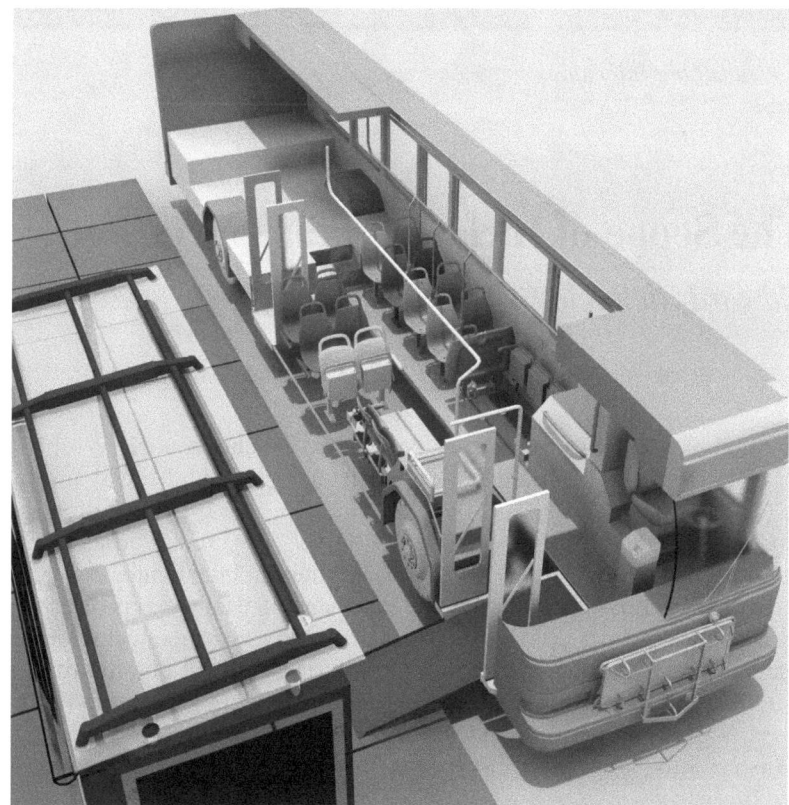

Figure 2.7 Image of virtual model of the simulated bus.
Source: IDeA Center

Photo of a transit bus interior.
Source: Author

3 The Scope of Inclusive Transportation

Edward Steinfeld

Overview

Transportation plays an important role in creating an inclusive society. It provides independent access to employment, education, health facilities, and social and recreational activities (IDRM, 2005, 2007a, 2007b). Without accessible transportation, people with disabilities are more likely to be excluded from services, social contact, and become stuck in a disability-poverty cycle (Roberts & Babinard, 2005; Venter et al., n.d.).

People with disabilities are not alone in experiencing exclusion as a result of limited transportation access. Any citizen who does not have a driver's license or access to an affordable automobile will encounter similar barriers in most communities as a result of limitations in public transportation services—infrequent service, routes that do not serve all parts of a community, and other service limitations. In many communities, public transportation is simply non-existent (Bailey, 2004). Thus, lack of service in itself is the major barrier to transportation access. Since poverty and disability rates are associated, communities with high poverty rates are likely to have higher disability rates, increasing their residents' dependency on public transportation. Finally, the most vulnerable, those who are not able to drive and also have incomes so low that they cannot afford an automobile, will not only be dependent on public transportation, but also dependent on accessible facilities, equipment, and services.

The concept of transportation dependency masks an important consideration in public transportation, the impact of improved services on the entire population of a region. In a free market economy, the mode of transportation that one chooses ideally would be market driven. But, in fact, as discussed in the last chapter, urban planning and transportation policy shape the choices that are available and the costs and benefits of each mode. Heavy investment and subsidy of highway systems, as opposed to public transportation, makes it difficult for public transportation to compete for "market share" with private automobile transportation. While a trend toward higher fuel costs and losing the ability to drive through aging or disability can lead to mode shift, improved service and improved service quality also contribute to the decision to use public transportation as opposed to private automobiles. Higher investments in public transportation and better use of existing resources to improve service availability and quality have benefits for all residents of a region by encouraging mode shift. These benefits include increased competitiveness in attracting businesses that depend on local transportation, reduced commuting times and costs for automobile commuters and public transportation riders, reduced air and water pollution, and reduced commuting costs.

In this chapter we will review the scope of accessible public transportation including access to services, operating policies, traditional accessibility provisions, and universal design approaches. The concept of universal design recognizes that the benefits of accessible transportation extend to a wider population than just people with disabilities. In public transportation, universal design can contribute to improving the quality of public transportation but also, as noted above, to benefits for the rest of the population by attracting riders who may otherwise travel in private automobiles.

Transportation Service and Policies

What constitutes an inclusive public transportation system? It is clear that the first criterion is that service is available. Availability is not simply a matter of providing service to a neighborhood. It includes service that provides access to the destinations that people need to reach (e.g., worksites, shopping, health care settings, education settings, recreation settings, etc.). The second criterion is that it is affordable due to the relationship between disability and low income and the inability of many people with disabilities to drive. The third is that the elements of a system must be usable for people who have limitations to mobility, perception, and cognition.

In the U.S., more than half a million people with disabilities cannot leave their homes because of transportation barriers (Bureau of Transportation Statistics, 2003). Four times as many people with disabilities as people with no disabilities lack suitable transportation options to meet their daily mobility needs (National Council on Disability, 2005). Public transportation problems are typically worse in rural areas, where approximately 17 percent of Americans reside (Johnson, 2006). Rural transportation infrastructure and service is often insufficient or ill-equipped to meet residents' needs. The rural population has higher percentages of people with disabilities (21.5 percent), elderly persons (13.8 percent), and people living in poverty (11.8 percent) than the population living in urban areas (Brown,

2008). The resources of rural communities are much more limited than in suburban or urban locations. For example, approximately 45 percent of rural elderly persons and 57 percent of rural poor persons do not have an automobile. Despite the obvious need, 38 percent of all rural residents live in areas without public transportation service (Dabson, Johnson, & Fluharty, 2011). It is therefore not surprising that a lack of access to transportation is one of the most frequently cited problems for rural residents (Spas & Seekins, 1998).

Travel patterns differ based on purpose of the trip. Most public transportation scheduling and route planning puts an emphasis on travel to and from work destinations. This focus optimizes services to those of working age, usually able bodied. There is research on the travel patterns of older people (Santos, McGuckin, Nakamoto, Gray, & Liss, 2011) but we have not been able to find origin and destination research specifically for people with disabilities. Such information would be very useful for transportation planners and operators, especially if it identified differences in trip patterns for this group compared to the general workforce and older people. Even without such research, we can make some pretty good assumptions about the travel patterns of people with disabilities. Those who work and are looking for work would presumably have the same patterns as other working commuters. Those who are not in the workforce would probably have the same patterns as students and retired people. Students travel from their homes to schools and universities and the surrounding neighborhoods during peak travel periods. Many older people tend to avoid traveling during peak commuting hours and at night. They prefer going out during daylight hours when there is less congestion; therefore, they tend to travel most frequently during the midday and on weekends (Alsnih & Hensher, 2003; Collia, Sharp, & Giesbrecht, 2003; Hess, 2009). Not surprisingly, older adults travel less frequently to places of work than younger adults. The spatial patterns of work have changed as urban development patterns have changed since the Second World War. In Chapter 2 we discussed these changes and the related research gaps.

Transit improvement districts are often designated development zones along major transit routes. Such districts often have incentives for developers to construct higher density housing. One such incentive, for example, is the removal or reduction of parking requirements. Some cities have adopted policies to encourage the development of accessible housing within these districts as well. There is no research that we found that identified the success of such incentives in improving transit access and other outcomes for people with disabilities.

In response to the affordability criteria, there is a widespread practice of giving people with disabilities, seniors, children, and students fare reductions. But, we found no research on best practices related to fare reductions. In our visits to transit agencies, we discovered that there are many different policies and procedures regarding fare reductions, sometimes confusing and burdensome to the rider, especially the first time user. In most systems in the U.S. one must prove eligibility for a fare reduction, which often requires an in-person interview. This could mean traveling to a special office and providing specific information (e.g., a Medicare card). Some agencies do not require an eligibility process and others require it for some groups and not others. In some systems, there are different fare reductions for different people (e.g., wheelchair users, visually impaired individuals, youths, and senior citizens). Each system usually has different ticketing options (e.g., monthly pass, daily ticket, one trip ticket, etc.). To complicate matters further, large urban areas often have many transportation providers, each with their own policies. The Bay Area has one of the more complex systems. A daily commuter to San Francisco may use a Caltrain commuter train to get into the city and then use the Muni light rail or bus to get to their worksite. A visitor to the city may also use ferries and even a cable car to visit tourist destinations. All of these services are operated by different agencies.

Complex policies make understanding fare payment difficult but they also have an impact on equipment design because each manufacturer markets their equipment to a broad range of systems and thus has to design equipment to accommodate a variety of policies. We found an example of the problems this can introduce in one large urban system, which recently upgraded all their fare gates to use with smart cards and trip tickets at a cost they reported as $40 million. Only one fare gate at each station was programmed for use with reduced fare cards and tickets. These gates were also the only wheelchair-accessible gates. Thus they had wider panels that were also heavier. The mechanism that retracted the panels was the same for all gates. Engineers expected the utilization of the "accessible" gates to be much lower than the others so they did not increase the durability of the mechanism to handle the heavier panels. Since the gates had to have signs to identify them as the path for those with reduced fare tickets, large signs were printed on them with the words "Reduced Fare." Presented with wider, more comfortable gates and the possibility of a reduced fare, riders pick these gates much more often than the others, thereby causing frequent equipment breakdowns due to durability problems (Figure 3.1).

Most of the research and practice in accessible transportation focuses on the environment of transit systems. In general, the focus is on meeting minimum design standards. But, meeting standards does not guarantee usability. Like other riders, people with disabilities need a system that is understandable and intuitive to use, safe and secure, and convenient. This includes the stations and stops, the rolling stock, the equipment like fare machines and gates, and information on schedules and routes—in other words, all physical and informational components.

Figure 3.1 Photo of fare gates in a major city's subway system. The wider "accessible" fare gate proved very popular with all riders.

Source: Author

Research on accessible public transportation has also identified the importance of the pedestrian environment to usability (Steinfeld & Seelman, 2011). The pedestrian environment is part of what transportation planners call the "last mile" problem or "first mile" problem—the difficulty of getting people and goods from a transport hub to their final destination or from their origin to the transport hub. This term is actually a figure of speech. The actual distance may be less than a mile or more than mile. There has been some research on this topic for general transit riders. Research on pedestrian activity finds that, as a rule of thumb, people will not walk more than 5 minutes to a destination. This is generally interpreted to be 1/4 mile (400 m) (Agrawal, Schlossberg, & Irvin, 2008; Southworth, 2005). Thus, urban planning guidelines for walkability recommend neighborhood design that provides facilities, like bus stops and stations, no more than 1/4 mile from any home. Research on older pedestrians and local mobility of people with disabilities indicates that they face significant challenges in the pedestrian environment (Kirchner, Gerber, & Smith, 2008; Koontz, Roche, Collinger, Cooper, & Boninger, 2009; Lobjois & Cavallo, 2009; Mollenkopf, Bass, Kaspar, Oswald, & Wahl, 2006; Wahl, Oswald, & Zimprich, 1999). There is no research to identify whether the 1/4 mile rule of thumb is applicable to people with disabilities.

Problems

- Limited public transportation services exist, especially in rural areas.
- Disparities in service availability for people with disabilities and older citizens.
- Complex and disparate fare reduction programs.
- The "last mile" problem.

Recommendations

- Explore the transportation needs of the most vulnerable populations since poor access to transportation is related to social isolation, lower community participation, and lower life satisfaction.
- Address the needs of the rural poor, many of whom do not have an automobile and live in places not served by public transportation at all.
- Consider how existing research (e.g., people will generally not walk to transportation stops further than 1/4 mile) applies to people with disabilities or older riders.

Directions for Future Work

- Information on best practices and identification of innovative solutions for providing public transportation in rural areas.
- Fare reduction practices and their impact on usability of systems need to be studied to identify the extent of the problem and best practices for implementation.
- Origin-destination research that includes a focus on disability and its relationship to age and income, employment status, and other factors. Such research would help to identify priorities and differences in needs.
- Outcomes-oriented research on the usability of transportation systems, including attention to both the information and physical environments.
- Research on the "last mile" problem to discover whether disability and age are modifying factors that should be applied to reduce the distance to transit stops for these populations.
- Research on the impact of transit improvement district incentives for accessible housing.

Progress in Providing Accessibility

Public transportation includes three basic types of service: fixed route, on-demand (e.g., taxis, jitneys, and car services), and special services (paratransit in the U.S.). On-demand service is a relatively new strategy in public transportation. Traditional systems included only fixed route services. Paratransit was developed to provide accessibility to those who could not use fixed route systems. On-demand service, usually provided by private operators licensed by public commissions at regulated market rate prices, provides a door-to-door service. However, on-demand services are now increasingly being viewed as an alternative to paratransit and fixed route services as well as a solution to the "last mile" problem for other riders who live beyond 1/4 mile from stops. Transportation services provided by private entities are covered by Title III of the ADA. There has been a great deal of controversy about accessibility to taxis in the U.S. The debate has focused on whether all taxis have to be accessible or just a portion of any fleet. The European experience with accessibility to taxis is much more extensive and documented in a recent research report (IRTU, 2007). Some cities, like London, have policies in place to convert the entire fleet to accessible vehicles. On-demand services can be implemented by public transit agencies to serve low-density areas and on "flex routes" in which vehicles deviate from a basic route to provide door-to-door service. They can also be used to solve the "last mile" problem, getting riders between the last station on a fixed route to their destination and vice versa.

Accessibility laws like the Architectural Barriers Act (1968), Section 504 of The Rehabilitation Act of 1973, The Air Carrier Access Act (1986), and the Americans with Disabilities Act (ADA) (1990) specify minimum requirements to ensure that public transportation does not discriminate against people with disabilities. In every system we visited, accessibility was a high priority, sometimes because of class action lawsuits against the providers. In new facilities and equipment, compliance with regulations seems to be good with respect to design. The challenge of adapting existing infrastructure is quite significant, however, especially deep tunnel stations without elevators and commuter rail with on grade platforms. In Tokyo, a city with a large number of deep tunnel stations, the transit agency launched a five-year effort to make as many stations as possible accessible by installing elevators and lifts. Within the five-year period, they were able to make 75 percent of the stations accessible. The last 25 percent are the ones where the cost and technical challenges are so great that "heroic" efforts have to be initiated with a much longer planning horizon. In the U.S. many older commuter rail stations are also historic buildings, which create opposition to improving accessibility or restrict options. For example, in the SEPTA system, there are older commuter rail systems with surface boarding. Higher platforms are needed to provide access to trains. But, building platforms would destroy the historic character of the stations. And, in some cases, there is simply not enough room to add a raised platform and ramp, thus requiring rebuilding the existing building and site or relocating the station to a nearby site.

Despite progress in design, there are still many problems related to services and operations. For example, both the literature and our research identified the problems of bus drivers who refuse to pick up wheelchair users and breakdowns in lifts and elevators that take a long time to be repaired. These problems can be addressed by introducing better training, better supervision, and improved maintenance practices. For example, one agency had a policy that a bus driver had to test out the operation of the lifts or ramps on their bus before starting their shift. But, they allowed the driver to do this outside the bus yard. As a consequence, there was no supervision of the testing procedure and many drivers did not do it. Another agency had the same basic policy but requires testing before leaving the yard where the procedure is under supervision. Some agencies have installed devices to provide real-time status of elevators so that broken elevators can be identified and repaired immediately. There are also factors beyond the control of the providers. For example, one agency had very old elevators but not the resources to replace them. Some replacement parts have to be machined since they are no longer produced. This took months to accomplish until they found an independent machine shop that was willing to do the work with a fast turnaround.

As noted above, meeting federal mandates alone is not likely to solve the problem of accessible transportation. In fact, some of the agencies we visited reported that a large proportion of people using paratransit on their systems are people with mental conditions, morbid obesity, and individuals who cannot get from their home to a transit stop. Accessibility standards for fixed route service do not adequately address the needs of these riders. The cost of paratransit service is becoming unsustainable as the number of older people in the population increases. The burden of paratransit can reduce service routes and threaten the affordability and quality of public transportation for everyone in a community, especially, as in the current economic climate, when federal subsidies are decreasing while ridership is increasing. Thus, public transit agencies are establishing stricter criteria for eligibility to paratransit services in response to the increased financial burden. This is a questionable policy within the context of the ADA and Section 504 but perhaps necessary for survival of the transit agencies. Many agencies are experimenting with incentives and a variety of alternatives like smaller fixed route vehicles, flex route, and on-demand services. In other countries, private taxi services are often used by transit agencies as an alternative to paratransit, graying the boundary between on-demand and special transportation. In the U.S., increased advocacy to make taxi fleets more accessible is making national headlines (Flegenheimer, 2012).

Ultimately, the success of accessible public transportation may hinge on its ability to draw riders, disabled or not, away from private automobiles and expensive paratransit. The former increases the efficiency of systems and allows them to expand their service area. The latter reduces costs per passenger. It can be argued that the lack of success transit providers have had in accomplishing both goals is due to their lack of a consumer-oriented philosophy. Most research on attracting riders has demonstrated that transit has to be more convenient, comfortable, safer, and more economical than private transportation before people will switch modes. Applying consumer-oriented strategies to public transportation can both increase usability for people with disabilities and also have great benefits for agencies by attracting more riders. One study in Canada, for example, found that only 25 percent of commuters liked their ride to work (Schwartz, 2011). A Harris Interactive poll commissioned by The Workforce Institute found that four percent (5 million) U.S. employees have called in sick because they could not manage their commute to work. Moreover, only 11 percent of workers used mass transit to get to work (Reuters, 2011). So, despite the generally positive view people have toward automobiles, there is potential for attracting riders if public transportation can be upgraded to provide an attractive alternative in comparison to private automobiles. As noted above, mode shift to public transportation also reduces demand for highways and if significant enough, could reduce highway infrastructure and maintenance costs as well as congestion and pollution. The public benefit goes well beyond the current riders of public transit.

Problems
- The high cost of paratransit service.
- The relative lack of accessibility in on-demand services like taxis.

- Difficulty making deep tunnel and surface boarding platforms in heavy rail systems accessible.
- Poor policies, training or compliance related to accessibility equipment breakdowns.
- Providing accessibility in older stations.
- Perception that accessibility benefits only people with disabilities.

Recommendations

- Address the many problems with service delivery related to training, management, and maintenance.
- Incorporate consumer-oriented strategies already identified as beneficial.
- Explore accessibility issues that extend beyond the current scope of standards.
- Utilize research from Europe that identified best practices in on-demand services.

Directions for Future Work

- Identifying the ridership profile of paratransit systems and the reasons for not utilizing fixed route services.
- Identifying problems related to renovating existing older stations and finding best practice solutions.
- Identifying problems related to operations including training, maintenance, and supervision needs.
- Research on the cost benefit of universal design in transportation systems.

Universal Design

Experience with accessibility laws led Ron Mace, Ruth Hall Lusher, and others (Bednar, 1977; Lusher & Mace, 1989; Welch, 1995) to recognize the need for a new approach to improving access to the environment, which they termed "universal design." The premise for this new approach was that the environment can be much more accessible than laws can realistically mandate on the basis of non-discrimination. Universal design offers a promising alternative to conventional accessible design in transportation, since it improves usability for everyone, not just people with disabilities (Danford & Maurer, 2005). "Universal design is an approach to design that incorporates products as well as building features which, to the greatest extent feasible, can be used by everyone" (Mace, 1985). The Principles of Universal Design were developed to create a conceptual framework for practical applications: (1) Equitable Use, (2) Flexibility in Use, (3) Simple and Intuitive, (4) Perceptible Information, (5) Tolerance for Error, (6) Low Physical Effort, and, (7) Size and Space for Approach and Use (Connell et al., 1997).

Universal design can be used to inform policy and intervene at the systems level to increase the potential for developing a better quality of life for a wide range of individuals (Russell, 1999; Stineman, Ross, Fiedler, Granger, & Maislin, 2003). Steinfeld and Maisel (2012) argue that universal design is best conceived as a process of continual improvement toward the ultimate goal of full inclusion. Experience with universal design in practice demonstrates a need to address contextual issues in practice. For example, a poorly funded rural transportation system may not be able to incorporate the same features and services as a large metropolitan system. Steinfeld and Maisel also point out that the definition above and the Principles are lacking attention to health and wellness issues and do not clearly address social participation goals. Further, they are not aligned with specific bodies of scientific knowledge and thus make it difficult to find and apply that knowledge to design practice.

Recently, they updated the definition of universal design to address these issues more explicitly (Steinfeld & Maisel, 2012):

> Universal design is a *process* that enables and empowers a diverse population by improving human performance, health and wellness, and social participation.

In short, universal design makes life easier, healthier, and friendlier. Their definition frames universal design as both an idealistic approach in the long term and a realistic approach in the short term. Moreover, unlike the original definition, they argue that universal design is equally useful in designing the virtual world of information and in delivery of services as it is in designing the physical environment. While they acknowledge that the Principles of Universal Design have played an important role in helping to clarify the concept of universal design, they offer eight Goals of Universal Design to address the limitations above and to provide further clarification of the concept:

1   *Body fit.* Accommodating a wide a range of body sizes and abilities.
2   *Comfort.* Keeping demands within desirable limits of body function.
3   *Awareness.* Ensuring that critical information for use is easily perceived.
4   *Understanding.* Making methods of operation and use intuitive, clear, and unambiguous.
5   *Wellness.* Contributing to health promotion, avoidance of disease, and prevention of injury.
6   *Social integration.* Treating all groups with dignity and respect.
7   *Personalization.* Incorporating opportunities for choice and the expression of individual preferences.
8   *Cultural appropriateness.* Respecting and reinforcing cultural values and the social and environmental context of any design project.

It is important to note that universal design does not substitute for assistive technology nor does it eliminate the need for regulations that define the legal baseline for minimum accessibility. But, conceiving universal design as a continual improvement process implies that regulatory efforts need to be complemented by additional policy initiatives, performance criteria for design, training, and management practices. Universal design reduces stigma by shifting the focus of service delivery from people with disabilities as a protected class to general improvements in services and design. In the process it leverages economies of scale for implementing innovative ideas.

The low-floor bus is a good example of the universal design approach applied to public transportation. The original approach to making buses accessible was to equip existing high-floor bus designs with lifts. Changing the bus chassis to one that was easier to make accessible with a ramp was a major breakthrough in accessibility. Furthermore, adding a kneeling feature reduced the first step, which was often very high, to one that is the same height as a typical stairway. Combined with a raised platform, often no higher than six inches, direct platform boarding can be achieved with little if any slope on the ramp. Ramps may not even be needed if docking systems can eliminate the horizontal gap. The low-floor model is now used in heavy rail, light rail, and streetcars as well as in buses. It has become the standard practice in the industry.

The low-floor vehicle has many advantages. All passengers can board at the same time in the same manner; the extra waiting time required to deploy lifts is eliminated and no passenger becomes the object of anger at delays in service. Ramps can now benefit many riders who had difficulty with stairs, not just people who use wheelchairs. The low floor reduces effort for all passengers and reduces boarding time in general. It is important to note that, as discussed later in this book, the current low-floor bus model does have its limitations and now that we understand what they are, an additional cycle of improvement is warranted. Furthermore, low-floor buses may not be the most appropriate solution for every transportation application. In Bus Rapid Transit (BRT), for example, industry experts point out that high-floor buses provide increased capacity and can be made accessible through platform boarding, increasing safety and convenience as well as overall system performance.

As the example above demonstrates, many other consumer groups stand to benefit from universal design and accessible public transportation, for example, children, parents with strollers, travelers with baggage, and bicycle riders, groups often overlooked by conventional practices. And, all people, at one time or another, are "disabled" by the demands of the environment when they are sick, injured, tired, in an unfamiliar place or carrying a heavy burden. Thus, adopting universal design approaches can increase the constituency and political support for making improvements in accessibility of systems.

The importance of accessible transportation in the lives of people with disabilities and seniors has increased the need for more innovative universal design solutions and better knowledge about how and where to apply universal design in public transportation. More information is needed about successful practices, especially those that can be implemented within the constraints of current funding levels. Beyond accessibility issues, universal design improves public transportation's viability as an alternative to private automobiles for all riders.

Despite the general favorable reception toward universal design, the concept has not yet been widely adopted in the transportation industry. Many providers simply do not know about universal design or believe that it will result in additional costs. There has not been any research to document examples of universal design and study the economic costs and benefits. The costs of not adopting universal design approaches should also be investigated. For example, the lack of a universal design approach to fare gates as noted above, was a costly decision for the agency. They must now renovate the new fare gates in every station in their system. The director of accessibility services said that they had decided to make all new fare gates and replacement gates accessible for wheelchair users and also programmed to accept reduced fare cards.

Identifying Best Practices and Lessons Learned

Finding good examples of accessibility and universal design in public transportation is not easy. Access Exchange, a not-for-profit established by Tom Rickert in San Francisco, publishes a newsletter with reports from all around the world. They have some regional correspondents that provide good insights and information on developments in their countries or regions. These reports have some details on design and service practices. The International Disability Rights Monitor (IDRM) series also provides some information on a global basis, with contributions from different countries. The IDRM reports focus on policy issues rather than details. The World Bank has been active in writing reports about the state of the art but these reports also tend to focus on policy rather than practices. The TRANSED conference, held on a bi-annual basis in different countries, provides a source of research papers and documentation on different practices, especially on the practices of the host country. The short format of the papers limits the level of detail available. In the last two conferences, much of the content has focused on accessible tourism rather than design and technology.

The TRB is one organization that has an ongoing program on accessible transportation. The conference is held every year in Washington, DC, and the proceedings are available online. This organization overlaps in participation with the TRANSED conference. A major problem with the current literature is the lack of good quality control on publications. There is no organization that has developed a systematic method for identifying, evaluating, and documenting examples of accessibility and universal design. Moreover, there is no peer-reviewed journal dedicated to the topic or one that has regular articles on it. Although TRB and TRANSED are peer-reviewed, there are no set standards for publication of examples from practice other than clarity and language. Most of the papers about practices document local examples without systematic evaluations using reliable methods. The RERC successfully demonstrated a method for systematic evaluation from a user perspective. This method could be adapted for this purpose.

From the sources available, we know that access to public transportation varies greatly around the world. Western Europe has some of the best examples of accessible public transportation, but serious problems remain in the region, especially in Eastern Europe (International Disability Rights Monitor, 2007b). One primary problem is the lack of effective policies that address the needs of seniors and people with disabilities. In London, studies revealed that only 10 percent of trains and 29 percent of buses met access standards (Kuneida & Roberts, 2006). Even where they exist, there is limited compliance with existing laws, especially in the developing world where a regulatory framework and funding are rarely put into place to implement legislation (Kuneida & Roberts, 2006; Roberts & Babinard, 2005). The benefits of universal design features are not well understood and thus many policy initiatives that could benefit all riders as well as transit system performance are not incorporated, for example, raised boarding platforms that reduce the need for stairs to access buses and reduce overall boarding times for all riders (Roberts & Babinard, 2005).

Unlike in urbanized and developed countries, low-income countries have very limited access to public transportation. The IDRM reports that, in the Americas, only 20 percent of the countries in a recent survey had an accessible bus transportation system although the U.S. and Canada reported accessible systems in all cities (International Disability Rights Monitor, 2004). Many transit providers, particularly in the developing countries, have implemented accessibility only partially, providing a limited number of vehicles equipped with lifts or ramps on each route, improving only key stations, or providing access only on new lines. While laudable as a start, ultimately, these efforts are not sufficient. To utilize a transportation system effectively, people need to have access to all vehicles and the full service area as well as the pedestrian environment (Iwarsson et al., 2000). Strategic planning is needed to establish a long-range planning process for improving accessibility throughout a system, including the pedestrian access to systems. In low-income countries, many alternative and informal types of transportation are emerging, such as jitneys and pedi-cabs (Cervero, 2000). The effectiveness of these systems and the potential for widespread adoption, even in high-income countries, has not been explored.

In new construction projects, the use of BRT is a major trend globally and especially in developing countries of Latin America and Asia. A BRT system is like a rail system with buses, with few stops. It often has stations and terminals to which local routes feed passengers. In high-level BRT systems, large buses run on dedicated bus-ways with a limited number of stops; raised platforms are used to load large numbers of riders quickly to high-floor buses, which have higher capacities than low-floor models. In low-level systems, BRT buses run in mixed traffic and load to slightly raised platforms (curb height). Low-floor buses can be used to provide access. Notable accessible BRT systems have been constructed in Curitiba, Bogota, Quito, Brisbane, Calgary, and Rio de Janeiro (Wright, 2004) (Figure 3.2). Despite extensive funding, no systematic research has been completed to study the impact or effectiveness of accessibility provisions in these massive projects.

Figure 3.2 BRT system in Rio de Janeiro, Brazil. (a) Bus and dedicated right-of-way, (b) boarding platform.
Source: (a) Photo courtesy of Tateyama, (b) Photo courtesy of the IDeA Center

Problems

* Adoption of accessible transportation practices is uneven globally, although many innovative approaches have been implemented.
* Practical solutions for developing countries are needed and need to be evaluated for effectiveness.
* Timely information on innovative solutions to accessible transportation problems developed outside of the U.S. are not easily accessible to the U.S. industry (e.g., policy makers, design professionals, operators, manufacturers). This is especially true for on-demand and flex-route systems.
* Quality control on practices disseminated is limited. At best, identification of a best practice is based on the opinion of one expert.
* Massive BRT projects have not been systematically evaluated for accessibility.
* The belief that universal design applications increase costs.

Recommendations

* Utilize existing research that documents the countries that have successful practices in place.
* Apply best practices already studied in Europe to the U.S. transportation industry.

Directions for Future Work

* Developing user-friendly versions of existing best practice research.
* A website with good and poor practices could improve dissemination, especially if well publicized.
* Developing cost-benefit data on the application of universal design, including benefits to the cost of operations and analysis of the costs of not adopting universal design features.

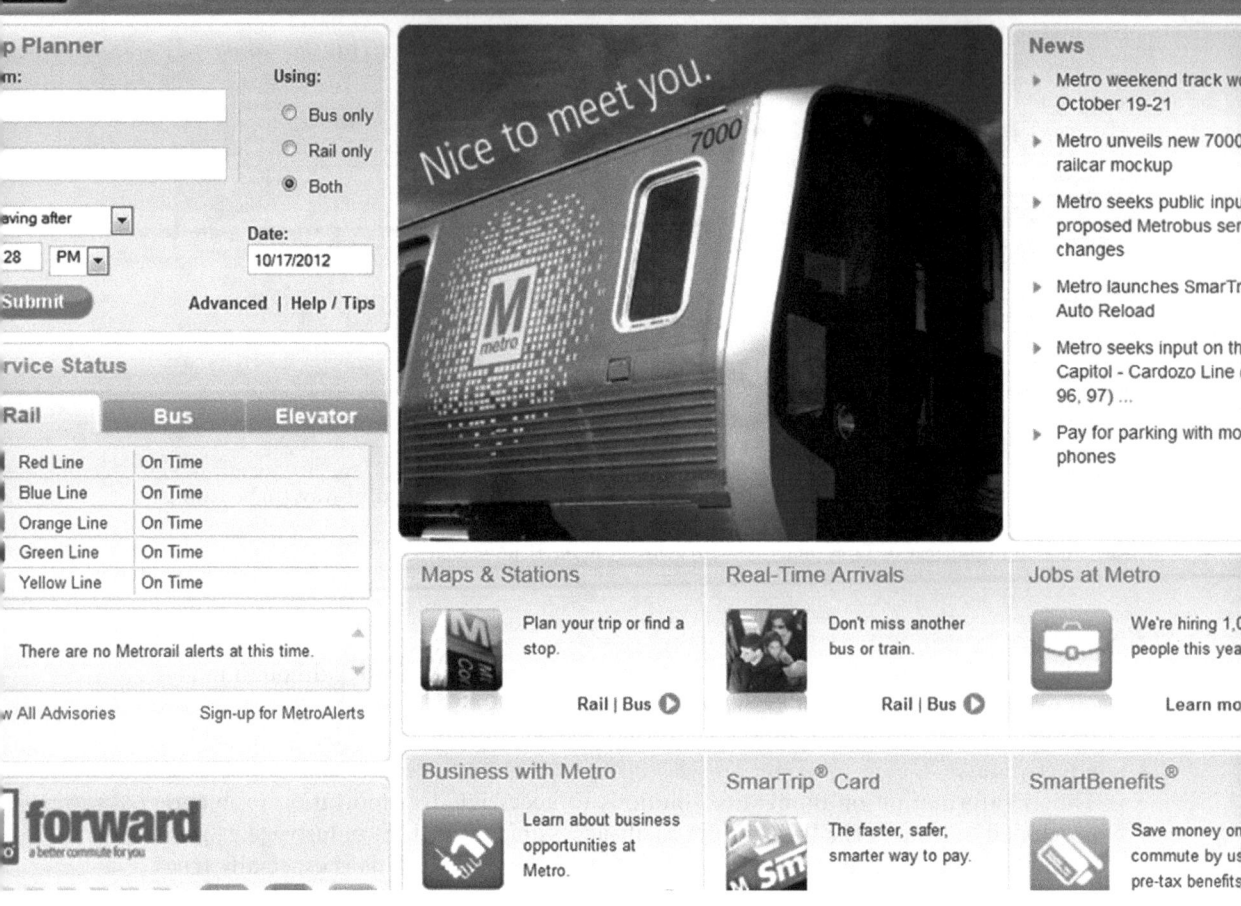

Photo of the Washington Metropolitan Area Transit Authority's website homepage.
Source: IDeA Center

4   Trip Planning and Rider Information

Aaron Steinfeld, Heamchand Subryan, and Ellen Ayoob

Overview

As is evident from the first few links in the Travel Chain, good rider information is important when planning and conducting a trip. Primary methods used for providing trip-planning information are printed media, help lines, websites, and smartphone applications. In general, all large transit systems provide trip planners and travel information on their websites as the primary outlet for rider information. Smartphone apps from transit providers are increasingly available to access this information but other apps also provide additional information sources. Printed media are often found at bus and rail stops, although the information provided is usually limited to the routes served by the stop and it is often permanently affixed to shelters, information boards, or signboards. Printed route maps and schedules are generally distributed only at major terminals although some agencies also provide information on vehicles (e.g., handouts for the routes served). Most large transit agencies also have some type of help line although the ease of using the service, the degree to which human operators are present, and the hours of operation vary considerably from agency to agency.

Traditional Information Channels

Paper schedules, brochures, and telephone-based customer service are still offered, but to reduce operation costs and improve sustainability transit agencies are seeking to abandon these information channels. The cost of printing thousands of paper documents that will become obsolete within a few months (due to permanent and temporary schedule changes) is not an efficient use of resources. Additionally, many paper schedules are left in distribution stands and never picked up by riders. At a time when all organizations are evaluating their impact on the environment, there is social and political pressure to reduce the use of paper.

Conversion to a digital mode of communications can also be motivated by accessibility. Personnel from one transit agency report that they moved away from paper schedules and postings at bus shelters over fears of being sued under the Americans with Disabilities Act (ADA) for not providing alternative modes of information for people with visual impairments. While there are limitations to paper schedules from an accessibility perspective, it should be noted that, compared to help lines and digital information accessed by smartphones, they are easier to use while on vehicles, reduce exposure to crime, and are attractive even during bad weather. Riders are more willing to risk water damage to a disposable paper schedule than their phone. Other weather factors can also suppress phone use. A student ethnography project conducted by our team found that riders use smartphones sparingly at bus stops during high-temperature weather. Other student work involving interviews with riders with disabilities revealed that participants would cut out and annotate paper schedules to make personalized booklets. This is, obviously, hard to do with other sources of information.

Help lines may be more environmentally sustainable but they are also costly to operate. Telephone call centers across the country are under pressure to control costs due to budget cuts and legacy costs (e.g., union contracts with mandated salary increases, pension costs, etc.). It is not uncommon for agencies to restrict call center hours, and, very long hold times are common due to reduced staffing. Yet, in direct opposition to these limitations in help center access, participants with disabilities in our research identified phone communications as their main mode of interaction with the transit service (Yoo, Zimmerman, Steinfeld, & Tomasic, 2010). They called to get schedule information, to plan trips, and to find out why a bus was late. With the exception of a few instances, they all seemed happy and comfortable with the phone interaction. They liked the fact that a person had heard their words. They mentioned their dislike for automated phone information systems and voicemail.

Some of the customer call center load can be ameliorated by the introduction of improved technology. For example, one participant with a disability in our research reported waiting 45 minutes on hold to complain about a driver (Yoo et al., 2010). They were forced to leave a message and never heard back. Interviews with transit agency personnel revealed that approximately 85 percent of their incoming calls were riders asking, "Where is my bus?" These were most frequently associated with buses being late, a problem that could be rectified with an AVL system. Shunting bus arrival queries to an AVL-backed automated system can free up call center capacity for true customer service problems. Unfortunately, real-time arrival data based on AVL is sometimes expensive to collect and disseminate. An analysis of deployed systems estimates approximate costs for a basic system at $16.5 million for a mid-sized city (i.e., 800 vehicle fleet; Parker, 2008). The high capital and operating expenses associated with proprietary real-time arrival data systems is the primary reason for lack of adoption cited by agencies interviewed by our team.

Many people dislike automated call systems due to their confusing, constrained decision tree architectures and general reliance on keypad actions. There is a preference toward "chat" style interactions (Yoo et al., 2010). Thankfully, there have been advances in natural language interaction, especially in the context of transit schedules. The Let's Go! project focuses on natural spoken dialog with computer systems and has used the transit information domain as a testbed for advancing the state of the art. As of 2010, the project had served over 130,000 calls over a five-year window (Black et al., 2010; DialRC, 2012; Let's Go!, 2012). The usability of such systems for people with speech, hearing, and communications impediments needs to be studied.

Web-based Information

Websites are dynamic and can easily be changed to reflect detours, temporary schedule changes, updated schedules, and other frequently changing information. For regular transit patrons, there is strong evidence that such websites are perceived to be useful and are used on a continuous basis (Gildea & Sheikh, 1996). Many agencies provide printable schedules on their website, thereby reducing the need to distribute printed schedules at stops and on buses. Riders can download and print their own schedules as desired.

Unfortunately, many transit agency websites have accessibility barriers. Using the World Wide Web Consortium(W3C) Web Accessibility Initiative protocol (WAI, 2012), we conducted a study of five representative sample screens each from 11 public transportation agencies of mid- to large cities, a total of 55 screens. The sample included the home screen and then comparable screens containing forms, trip planners, PDF files, and images. The study included both automated and manual checks of the sample screens. Because websites are often updated frequently, we conducted three separate automated checks at different times during the year. The amount of errors found on the first check changed on subsequent checks. Although some of these errors were fixed on one screen, the same errors often emerged on another screen. Therefore, it is important to consider accessibility each time the site is altered.

Both the initial review process and final results revealed common errors across sites. The top five common errors found included missing web form label, missing alt text, empty link, orphaned web form label, and linked image missing alt text. Of the five, the main problem was missing web form labels (e.g., labels for entry fields in trip planners or contact forms). The prevalence of this error has likely increased since the study because more transit agencies are introducing interactive form-based travel planners. Another major issue is missing alternative (alt) text that provides textual description for non-text content such as images. Alt text is the most basic and easiest accessibility improvement a site can make. At the time of this analysis, the Washington DC Metro website had the best accessibility of those examined. We highlight it here so developers have a best practice example to inform their efforts.

Transit website developers should be vigilant about accessibility when implementing a new website or updating an existing one. A filter is a quick, inexpensive way to determine if a recent change altered accessibility or introduced a potential problem that needs investigating. A web based filter like WAVE provides a toolbar that can be installed on various browsers to check for accessibility and provides links to WebAim.org, a comprehensive resource for accessibility related issues.

Good, easy to implement accessibility approaches have been available to website developers for years. There are standards, validation tools, and numerous instructional materials (WAI, 2012) yet developers are still slow to properly implement accessibility to websites. Researchers at IBM have attempted to bypass developer intervention by leveraging social networking technologies to crowdsource website accessibility. Their Social Accessibility Project developed a web browser plug-in that allows end users to flag problem pages, provides tools for sighted users to voluntarily add a markup layer, and then fuses the results to generate an accessible version of the page (Sato, Takagi, Kobayashi, Kawanaka, & Asakawa, 2010). This approach has promise but has not been widely adopted. It would be interesting to see how techniques developed through this work could allow non-programmers within agencies to refine and improve the agency's website behind the scenes.

Often riders do not know what accessible features are provided by their transit systems. A supplemental study was conducted to find out whether or not the 11 sites evaluated also included readily available information on their respective websites about accessible features provided to riders. Accessible information provided to disabled riders regarding stations, vehicles, and schedules should be prominently displayed on all transit websites. However, most of the transit sites evaluated buried this information in sub-pages that were hard to find. All of the sites provided generic information regarding system access for disabled riders such as, "all buses are equipped with ramps or lifts" or "paratransit service provided to eligible persons with disabilities who are, unable to board, ride or disembark from an accessible vehicle." Within the accessible information page, each transit site evaluated stated that all of their buses either had a lift or ramp with priority seating for riders with disabilities. However, information highlighting accessible features or routes within a specific bus station was not provided.

Trip planners are an easy way for riders to figure the best and fastest route for them to take. All of the sites examined had standalone trip planners, but some include Google Transit as an alternative. Of the 11 sites evaluated, six of them provided an option to select an accessible route when planning a trip. The validity of the actual accessible routes provided by the trip planner was not investigated. Additionally, all of the websites evaluated included pointers to some sort of smartphone application. Some of these smartphone applications were provided by the agencies themselves, while others referred users to applications developed by third parties.

Infrastructure-based Information

Obtaining accurate scheduling, route information, and real-time information on estimated arrival times is essential to provide a positive user experience (Norman, 2007). This data is particularly important to people with disabilities. For example, they may be more vulnerable to exposure in severe climates or have medical needs that require attention on a timely basis. Having information on breakdowns and delays can be critical for their health, comfort, and job performance.

But, even when such information is provided, it is often not in an accessible form. Providing fully accessible systems at each stop (e.g., text and audible schedule information) is costly and is usually reserved for major transit terminals. In fact, as noted above, the ADA requirements for accessible information may inadvertently be driving some transportation providers to reduce the amount of information they provide in general because it is so expensive to provide it in more than one sensory modality at each stop (Talbot, personal communication, 2008).

Travel information provided via text messages and infrastructure labels is an option (e.g., RouteShout, 2012). Riders text a bus stop label and receive upcoming bus times in

Figure 4.1 A multi-language information display inside a Japanese subway car.
Source: Aaron Steinfeld

response. This has value to people who cannot use voice interaction but riders may encounter challenges when attempting to identify their stop name. For example, stop labels may be in small print, mounted high in the air or visible only from the street side, and lack Braille alternatives. Some systems do not include infrastructure labels and require the rider to learn a special syntax. Near Field Communication (NFC) tags are a possible solution since they can be used to automatically trigger access to mobile websites on smartphones. Chapter 7 has more details on NFC and the integration of location-specific information.

Infrastructure within a vehicle is also a source of information (see Figure 4.1). Printed route maps are common on trains and dynamic message signs are also becoming more prevalent. Some vehicles now have small flat-screen displays for advertisements and trip information. Some service providers have employed novel approaches to improve accessibility within their vehicles. For example, Indian Trails recently installed hearing loops within their vehicles for hearing aid and cochlear implant users (Indian Trails, 2012). Another example is the Lausanne Metro, which plays meaningful sounds within vehicles approaching specific stations (e.g., water sounds when approaching the station near a fountain, etc.). This universal design approach has value to people who are blind, have a cognitive disability, or do not speak the local language.

Smartphone Apps

In recent years there has been an explosion of smartphone apps targeted at transit riders. This is in response to trends showing two-thirds of the U.S. population and three-quarters of U.S. teens have access to smartphones (Lenhart, 2015; Smith, 2015). Most smartphone apps provide schedule information and/or AVL-based real-time arrival information as well as trip planning, and detour notification. Some of these apps are associated with research projects and have uncovered interesting results.

OneBusAway (Ferris et al., 2010a), developed at the University of Washington, provides web and mobile access to real-time arrival information it gathers from the transit agency. Real-time information, as opposed to scheduled information, is broadly desired by transit riders and has significantly increased ridership for specific routes (Casey, 2003). Evaluations of OneBusAway showed that mobile access to real-time arrival information decreased uncertainty and increased users' perceptions of flexibility and safety (Ferris, Watkins, & Borning, 2010b). The authors also focused much of their field research on real-time arrival information through the Tiramisu system (Steinfeld, Grimble, Steinfeld, Rao, & Tran, 2012; Tomasic et al., 2015; Zimmerman et al., 2011). Since this system focuses heavily on social computing solutions to service delivery, detail is deferred to Chapter 8.

There are efforts to provide information to riders using mobile devices designed specifically for people with disabilities. For example, the Travel Assistance Device (TAD) tested in Tampa, Florida, directly supported riders with cognitive disabilities (Barbeau, Georggi, & Winters, 2010; Winters, Barbeau, & Georggi, 2010). In addition to TAD, a number of groups have explored delivery of transit information via mobile devices (e.g., Biagioni, Agresta, Gerlich, & Eriksson, 2009; Li & Willis, 2006; Masood & Nicholas, 2003) and other approaches have been used to support overall system navigation for riders with cognitive disabilities (Repenning & Ioannidou, 2006).

The TAD warrants specific attention for several reasons (Barbeau et al., 2010; Winters et al., 2010). The system allowed caregivers or personal care assistants to program in routes via websites for riders with disabilities. The system then tracked the users and provides prompts regarding which bus to board and when to get off. The rider's progress was also checked and alerts are issued to the caregiver if needed.

Over time, some of the text messaging systems migrated to smartphone apps (e.g., RouteShout, Roadify). Other apps began specifically as smartphone apps, either as multi-market solutions or rebranded versions of a common base application for a specific transit agency. Some apps have also begun to integrate Twitter and other social web content. Unfortunately, many apps and smartphone-friendly websites are not accessible, including those developed by transit agencies (NCAM, 2012). Given the problems noted above on agency websites, this is not entirely surprising. Furthermore, very few offer trip-planning options designed to support the needs of users with disabilities. For example, HopStop, a now defunct commercial transit planning website and smartphone app, provided a "Wheelchair Accessible/Stroller Friendly Route" option during trip planning.

There is progress toward broader incorporation of accessibility details into transit trip planning. The non-profit OpenTripPlanner team (OpenPlans, 2012) began to incorporate wheelchair accessibility as an option but, as of this writing, this feature does not work well. The NavPal (2012) team has been adapting robot navigation planning algorithms to handle the "fuzzy tradeoffs" needed by people with disabilities. Traditional navigation planning systems use "hard" constraints to generate routes (e.g., sidewalks only) but in robotics the constraints are often based on context (e.g., don't drive over dirt roads). This contextual approach is very important for people with disabilities. For example, a blind user may prefer intersections with traffic controls like signals but will tolerate intersections without controls if it means saving a lot of time.

Problems

- There are significant economic pressures for agencies to reduce printed and phone-based customer service information sources.
- Environmental sustainability and the ADA are producing pressure to limit, or eliminate, paper schedules.
- Phones are becoming increasingly useful for transit information yet they are susceptible to theft and weather damage.
- Information relevant to people with disabilities, like elevator and escalator status, is often missing in current information sources.
- Trip-planning algorithms and tools often lack options and information relevant to people with disabilities.

Recommendations

- If printed schedules and route maps are being phased out, consider offering printable hand schedules and route maps on a website.
- Improve the accessibility of websites and smartphone apps and regularly monitor accessibility as software changes.
- Incorporate research advances on natural language into automated telephone systems to support more chat-like interaction.
- Provide data to third-party developers so users can download the app which best suits their needs and abilities.

Directions for Future Work

- Data sources should be improved to include information relevant to people with disabilities.
- There is a need to develop methods and practices that support rapid information acquisition by riders before and during transit use, for all weather conditions.
- Methods for personalization of smartphone-based information need further development.
- Advances on chat style interaction that is usable by people who have hearing and communications disabilities are needed.
- Transportation standards and current industry practices should be adapted to incorporate universal design to promote innovation and also to ensure cross-disability accessibility.

Photo of a transit stop that includes a hybrid transit bus, bus stop signage, and a bus shelter.
Source: Author

5   The Built Environment

Edward Steinfeld

Overview

The built environment plays a very important role in the accessibility of public transportation systems. The built environment includes (1) pedestrian paths to stops and stations, (2) local stops, and (3) stations and terminals. The research on this topic is limited. There are some research studies of bus stops and shelters and many manuals that transportation agencies use to plan stops. There is practice-based information highlighting the importance of the pedestrian environment. These materials, information gleaned from agency visits, and presentations at TRB conferences informed this chapter.

Pedestrian Environment

Although it is a critical aspect of the Travel Chain, the design and maintenance of pedestrian environments outside the property of the transportation provider is generally the domain of public works departments and individual property owners rather than transportation providers. Close coordination between transportation providers and public works departments is therefore an important aspect of accessible transportation.

The pedestrian path system connects origins and destinations to the point of entry to the transportation system. If an individual traveler cannot access the pedestrian path system, then they cannot even start a trip, and if they arrive at a station or stop and cannot use that path system to reach their final destination, then they cannot complete their trip. Transportation providers have an obligation to ensure that the path system is accessible from their stops and stations. There are four important goals to reach in accomplishing this: accessibility, convenience, security, and safety.

Transit Stops

The location of transportation stops plays an important role in making a convenient connection to the pedestrian path system and providing security. Transportation providers should coordinate stops and stations with planning authorities and private landowners so that they can be located adjacent to major retail centers and worksites (Maynes Associates, 2009). Entries and exits to the stops and stations should be located and designed to reduce conflicts with automobile transportation and provide the most convenient and direct access to pedestrian path systems (Maynes Associates, 2009). These connections should be in high-use areas and have good lighting and views to and from adjacent activity locations (Newman, 1996). A good planning process would also explore the opportunities for commercial development and recreation amenities at the stops and stations (Project for Public Spaces, 1997). This usually requires the participation of municipal planning authorities, local residents, and other stakeholders in addition to the landowners since changes in zoning and waivers from land use regulations may be needed to provide the best solution (Maynes Associates, 2009). The Transect and the Smart Code can be used to provide ideas and guidance for appropriate development. A good example of careful planning is the Orange Line developed by LA Metro on an old railroad right-of-way. The land not used by the Metro was dedicated to a public park and recreation trail, generating increased pedestrian traffic in the vicinity of stations. Stations were located at the end of blocks so that they are visible to passing traffic on cross streets and the entry forms an activity node of their own, enhancing the safety of the sidewalks themselves. Park-and-ride lots and drop-offs were located away from the public sidewalks but adjacent to the stations to reduce conflicts between vehicles and pedestrians. A mall is a major destination along the Orange Line. A cross street bounding one side of the mall and directly across from a major entry was reconfigured to provide bus loading areas out of the way of traffic (Figure 5.1).

Bus routes generally run on public rights-of-way owned by municipal authorities. Although it is much easier to locate a stop on property controlled by the provider or a municipal authority, the Travel Chain concept highlights the importance of locating stops where they provide the best connection to the pedestrian environment, not only where they are convenient for the agency. In urban areas, bus stops directly on public sidewalks are generally the rule, but in suburban and rural locations, privately owned shopping plazas and malls are now the key employment and retail destinations. They often do not have sidewalks adjoining their properties, and large parking lots provide challenging conditions

Figure 5.1 BRT line in Los Angeles, CA, showing relationship to street, park, and bike path.
Source: Author

for pedestrians from roadside stops to the building entries. Ideally, the transit route should run into the private properties, and the stop should be located adjacent to building entries. Sometimes, topography or congestion in existing plazas and complexes precludes running the route to the buildings. In new construction, however, planning approvals can be contingent on provisions for transit access. Where this may not be possible, property owners should be encouraged to provide an accessible and safe pedestrian connection to the stop. Montgomery County's transportation agency has worked with plaza and center owners to coordinate stop locations and make good pedestrian connections to stores (Nabors, 2009). The parts of these walkways in public rights-of-way are financed, sometimes partially, by the agencies themselves to offer an incentive to the owner to provide the pathway.

Another important issue is protecting riders from traffic as they cross streets to reach bus stops or their destinations. Stops are generally located near corners to provide riders convenient access to pedestrian crossings with traffic controls. Where there are no marked crossings or signals at desirable stops, transit agencies can work with traffic engineering departments to provide them if pedestrian traffic is expected to be significant when a bus stop is constructed. At busy intersections, moving the stop out of the traffic lane with turnoffs can benefit pedestrians by reducing frustration on the part of drivers and increasing visibility to traffic. Montgomery County has also experimented with stops along busy roadways that are designed so that buses block traffic entirely while passengers are being loaded and unloaded, which gives the riders opportunities to cross the street safely. In these stops, traffic-calming features are provided in conjunction with bus stops. For example, a center island and bus bulb can reduce the width of the traffic lane to the width needed by the bus only. Riders disembark, cross the lane in front of the bus, and wait in the island until they can cross the opposite lane safely rather than crossing two lanes at once.

Bus shelters have complex requirements to ensure accessibility, convenience, and safety (Easter Seals Project ACTION, 2006, 2009). The highest-priority issues are accessibility for wheelchair use, visibility, climate protection, seating, scheduling, and route information. Although the Americans with Disabilities Act (ADA) has minimum requirements for wheelchair clearances, recent research demonstrates that it is advisable to increase the sizes required because there are many people not accommodated by those minimum standards (Steinfeld et al., 2010a, 2010b). Visibility is a key design determinant for security. Riders want to be visible and well lit to feel safe. They also want to see the approaching vehicles from the shelter and often will not use the shelter if such visibility is not available. Transparency and good illumination also are deterrents to criminal behavior and loitering. Although in cold climates it is desirable to have good wind protection, riders do not like to feel trapped in shelters with only one narrow entry and walls that reach the ground also collect trash and dirt and homeless people. They are also more difficult to clean (Figures 5.2 and 5.3). Seating is an important feature for all travelers, especially when long waits are common.

Although route maps and schedules are conventional elements of shelters and open stops, real-time information has been shown to reduce perceived waiting times (Dziekan & Kottenhoff, 2007; Schweiger, 2003). All stops should have curb ramps to ensure that passengers in wheeled mobility devices can board even if a bus has to stop far from the curb due to parked vehicles encroaching on the boarding area.

Figure 5.2 Although this bus stop is accessible by ramp, the entry to the shelter is too narrow to be accessible for wheelchair users and the shelter also blocks access to most of the stop. A bus would have to pull up in front of the shelter to load a wheelchair user.

Source: Author

Figure 5.3 This windscreen at a large bus shelter in a cold climate provides protection on all sides. Two entries improve access and security. The entries and the interior are wide enough for wheelchair access. The tinted glass reduces heat gain in summer and increases visibility of the stop from a distance.

Source: Author

Stations and Terminals

Layout: In general, stations and terminals have the same requirements with the exception that small stations may not present major wayfinding challenges. The sheer size of large terminals can create a significant physical burden on many travelers. Large terminals require careful analysis of options to reduce unnecessary travel distances. Automated transportation options like moving walkways and internal transit are valuable features of contemporary terminals. The wider range of passengers these devices can accommodate, the less special transport like people movers and carts will be needed. People movers and carts that share space with pedestrians complicate the terminal environment, increase the noise level, and often create safety hazards (Steinfeld, 2001).

Most travelers understand the common spatial organization of terminals—entry area, ticketing area, main circulation halls, departure gate areas, arrival gate areas, baggage claim (in airports and cruise ship ports), and exits. Terminal floor plans generally channel travelers through the terminals especially at security checkpoints, but in large terminals, there are often many choices to be made on either side of secure perimeters. A shallow spatial organization will reduce the number of decision points and thus reduce errors in wayfinding.

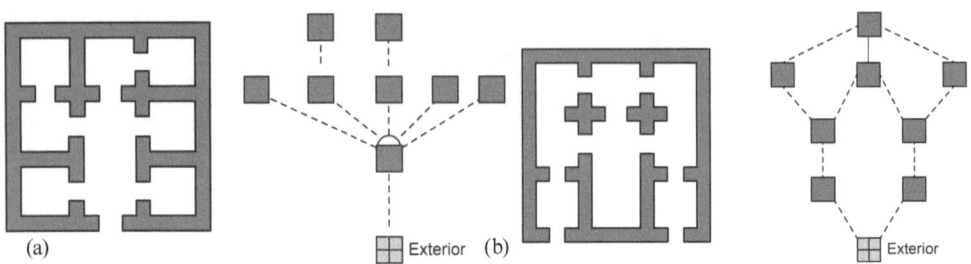

Figure 5.4 Diagrams of (a) "tree-like" and (b) "ring-like" designs.
Source: IDeA Center

In a tree-like terminal organization, travelers must make a series of decisions as they proceed deeper into the building. If they make a mistake, it becomes hard for them to understand the whole and it can take much effort and time to recover from the wrong choice and get on the right path again (Steinfeld, 2001). Changing vehicles while carrying baggage also requires a lot of effort in a tree-like plan, particularly if the arrival and departure gates are at the far ends of different branches. Although the overall system is easy to comprehend, actual experience of the plan is not the same as viewing a diagram of the whole due to limited visual access in the long "branches." A "ring-like" design has a much shallower spatial syntax. All the gates in each section are relatively close to one another and visible to the traveler from a central space. Transit and pedestrian links can connect the rings (Figure 5.4).

Signs: Identification information and directional information are needed in terminals. The former includes identification of key locations like ticketing counters, security checkpoints, information services, and gates. This information is typically provided by static visual signs. Design criteria for visual signs include (1) fonts large enough to be seen at the distance where people first encounter them; (2) contrasting text and background; and, (3) located overhead, the most visible position.

In public buildings, signs identifying rooms must be both tactile and visual, although only the room numbers need to be tactile. However, in transportation terminals, the effectiveness of tactile signs for room identification is questionable. Not only would it be difficult to find them in large congested spaces, but also a safety hazard on boarding platforms.

Increasingly, dynamic signs are being implemented in all types of terminals that provide more useful and specific information than a static gate number. These signs provide information on routes served, destinations, and scheduled departures and arrivals. They can also provide estimated time of arrivals. Dynamic displays should be protected from glare, have a resolution that is high enough for the distance viewed, and be free from color distortion and flickering. They should present new information at a rate within the abilities of people with sensory and cognitive limitations to perceive and process it (Figure 5.5). These issues must be investigated during building design because unexpected ambient conditions, for example, direct sunlight penetrating from clerestory windows, can make it difficult if not impossible to read the signs. In buildings with a lot of natural light, daily, seasonal, and weather-related variations in the quantity and quality of light need to be anticipated.

Using internationally recognized ideograms for signs used to identify important features or amenities (e.g., restrooms, exits, and information kiosks), is a major aid to the travelers who have sight but cannot read the local languages; they also reduce the amount

Figure 5.5 Real-time information provided on large, easy-to-read dynamic sign display.
Source: IDeA Center

of information that needs to be displayed by condensing words into symbols. Tactile cues and guide strips are useful for helping people with visual impairments find important destinations (Christophersen, 2002). These devices could be designed to benefit all travelers. They should be limited to providing guides to key destinations like platforms and restrooms. Too many guide strips can become very confusing, and discrimination among many types of tactile codes used for different destinations may be extremely difficult if not impractical.

Ideally, a Global Positioning System (GPS) orientation and wayfinding system could provide assistance in navigation for all travelers using both audible instructions and visual displays. However, GPS does not work indoors and does not have the accuracy to find destinations like gates or even boarding areas in terminals. Relying on GPS systems near a boarding platform could be fatal. Remote Infrared Audible Signs (RIASs) or "Talking Signs"® are a useful technology that overcomes these problems (Crandall, Brabyn, Bentzen, & Myers, 1999; Golledge, Marston, & Costanzo, 1998). These devices include transmitters at destinations that send out an infrared signal and receivers, which individuals hold to scan for the location of the devices. If the user deviates from the path, they lose the signal. And, as the user gets closer to the destination, the RIAS can provide more detail to guide the user very precisely. The conceptual model used in the device is much simpler than a GPS. It is like a homing beacon where the signal gets stronger as one gets closer, providing continuous feedback to help stay on the route. However, where there are lots of crowds, it may be difficult to pick up the infrared signal, which depends on line of sight, unless they are spaced close together and the ceiling is high enough to get good coverage of the path from above. The most important limitations of RIAS are that one needs to have a special receiver to use them; each signal must be hard wired, maintenance is required to keep the infrared light and lenses clean, and they do not provide dynamic information like that provided on LCD display panels.

Tactile maps can be provided by transit agencies to provide detailed information about stations. The Bay Area Rapid Transit (BART) system in the San Francisco Bay Area is experimenting with tactile maps that can be "read" with both fingers and smart pens. If more information is needed than a symbol or Braille can provide, the user gets a detailed audible description of the feature and location by touching a smart pen to the map. This type of map can provide complex and very specific information to help riders with visual impairments to understand the details of complex stations, and they also can aid other riders who may have difficulty understanding spatial relationships (Figure 5.6). Other technologies are under development to provide navigation assistance indoors using smartphones that could have significant advantages (e.g., NavPal, 2012).

In transportation environments, signs should be provided in more than one language. In the U.S., this would include English and Spanish, although there may be a need for more languages in certain neighborhoods. This practice acknowledges that the country is increasingly multilingual and that international tourists and business travelers from other countries are using public transportation. It also helps to create a positive emotional response. In large and complex transportation terminals, human beings, like information clerks, can provide personalized information on gates and schedules. Where the small size of a terminal does not justify clerks, remote human or automated help

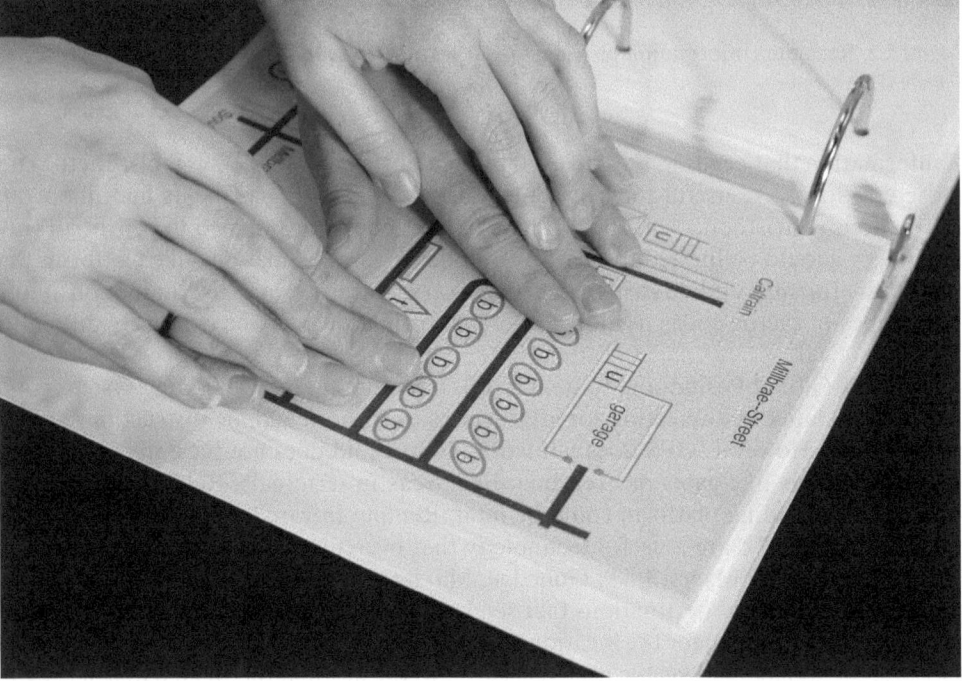

Figure 5.6 BART tactile map.
Source: Touch Graphics, Inc

that can be provided through telephone hot lines to a central information service is a good alternative. Many systems have free telephones that provide schedule information at each transit. The Japan Tourist Office offers a toll-free hotline in foreign languages that is available 24/7 throughout the country. Another option is an automated information kiosk with a connection to remote human or automated help. All these approaches can provide very useful assistance for all travelers but particularly people with visual impairments who cannot use dynamic signs.

Time-sensitive Information

Key sources of information in transportation environments are public address announcements, message boards, and video/computer monitors that display arrival and departure information as it becomes available. These systems often create barriers for people with sensory impairments. For example, message boards do not always have the same late breaking information being provided by public address systems. The message boards and monitors may not scroll and update fast enough to keep up with recent announcements. People with hearing impairments may miss information that is not available on monitors or message boards, like reasons for delays and track or gate changes. This is particularly a problem when using restrooms and waiting at gates in airports where flight information is not made available except through public address systems.

The advent of smartphones and wireless Internet access provides a technology for notifying individuals with hearing and visual impairments about changes in schedules in large terminals. Personalized "software agents" that collect specific information and notify the users automatically could be harnessed to support situational awareness. As noted above, airlines have already deployed automated texting systems to accomplish this goal, but a different system would be needed where the current riders are not known to the providers because they don't have to make reservations, e.g., commuter rail stations. A less advanced technology is real-time captioning of announcements at gates, platforms, and bus stops. Such a system would not reach individuals who were out of sight of the information (e.g., in the restroom).

Information Kiosks and Fare Systems

Information kiosks and fare systems provided at transit stations are required to be accessible by accessibility regulations, but compliance, particularly with interfaces, is uneven. Fortunately for developers, there are well-established guidelines and approaches for making such systems accessible (e.g., Trace Center, 2000). The problem at this point stems from executing best practices. There are good examples of specialized kiosks targeted at specific user populations. An example is kiosks for people who are blind with the combination of tactile information and interactive audio (Touch Graphics, 2008). Traditional tactile maps and three-dimensional models can be used effectively as well, but applications are limited, at the present time, to features in the environment that are permanent. New technologies are being introduced that combine three-dimensional models with digital information (see Figure 5.7) (Subryan et al., 2012). This technology has not been introduced into the transportation environment as yet. Remote human help on demand is also common and can be found in various forms. This can be the classic white courtesy phone or help buttons with speakers.

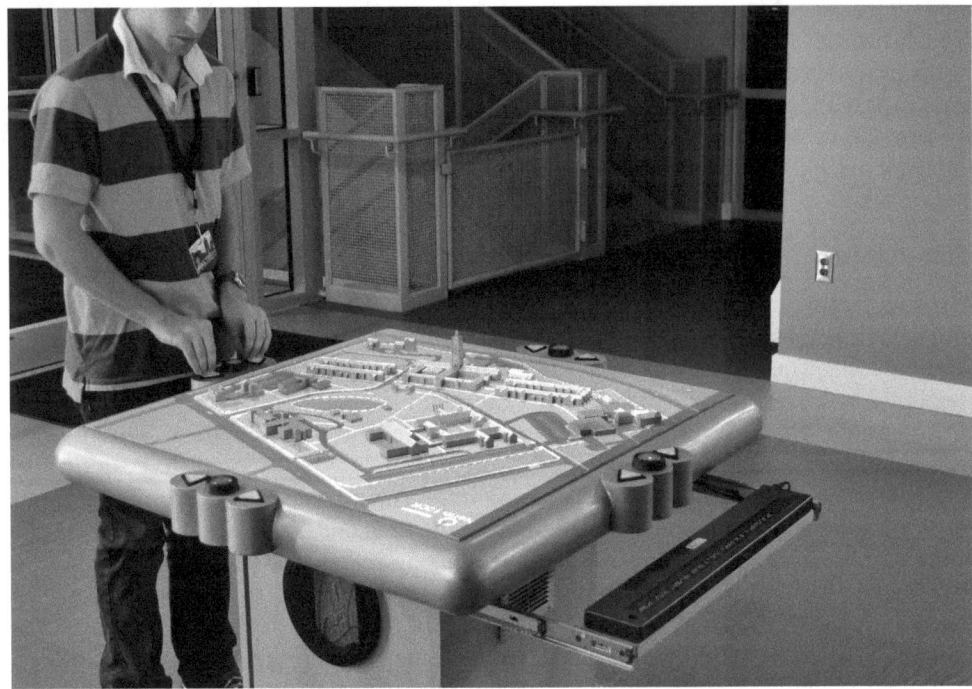

Figure 5.7 Multisensory Interactive Touch Model of a campus that provides tactile and audible feedback as the user touches different buildings and streets.
Source: IDeA Center

Level Changes

Level changes are often necessary in large terminals to efficiently process arriving and departing passengers and their baggage. Terminals that serve intersecting rail lines must have more than one level in order to allow one line to pass over the other. As in any multilevel structure, transportation terminals use elevators, ramps, and stairs to accommodate level changes. Even in small buildings like subway and elevated stations, the boarding platforms are often above or below the ground, out of the way of surface traffic. Where there are no boarding platforms, negotiating access to a vehicle is the most significant problem for access for people who have mobility limitations, including those who can walk but have limited strength, limited range of motion in lower extremities, or low stamina. Riders and system operators often prefer ramp access to elevators. The Southeastern Pennsylvania Transportation Authority (SEPTA) system engages patrons with disabilities in making decisions about accessibility strategies for improving existing stations. Their staff reported that ramps were preferred to elevators both for security reasons and also because there is no chance of a breakdown (Figure 5.8).

People with mobility impairments should not be forced to rely on assistance to negotiate changes in level. Forced dependency results in inconvenience, embarrassment, anger, and exposure to injury by poorly trained attendants (Kawauchi, 1999). This is especially the case when people who use wheelchairs must use an entirely different circulation path than other travelers. In new construction, providing accessible access to vehicles should be a primary consideration in the early stages of building design, in order to reduce the need for duplicate elevators or ramps (Steinfeld, 2010). From the perspective

Figure 5.8 An elevated commuter rail station in the SEPTA system was equipped with a long ramp rather than an elevator based on the preferences of riders with disabilities who were consulted in its design.
Source: Author

of management, universal design reduces the need for special training for all new staff, reduces occupational injuries related to lifting, and improves productivity because workers do not have to leave their posts as often to assist an individual traveler.

Vehicle Loading

One of the most difficult issues in universal design of transportation systems is accommodating level changes to board and exit vehicles. Two basic accommodation strategies can be used separately or in tandem. The first focuses on design of the vehicle and the second on design of the transit station or terminal. In addition to accommodating the level change, there is also a need to protect people from falls and other safety hazards in the loading area especially when there is a loading platform. A uniform design approach throughout a system eliminates surprises for the traveler, especially travelers who have visual impairments.

In new transportation systems for inter-urban and commuter rail, loading platforms should be at the same level as the vehicle floor. The gap between the platform and the vehicle should be eliminated or kept as narrow as possible to keep the front wheels of wheelchairs, walking aids, and people from falling into the gap. Consistency across the system and over time is important, so standards have to be established for both rolling stock and terminal construction. In existing systems where there are inconsistencies

Figure 5.9 Solutions for accommodating horizontal and vertical gaps: (a) portable ramp and
(b) folding ramp.
Source: IDeA Center

between levels, vehicles will have to be fitted with lifts or small ramps. The latter is preferable since all passengers can use them. But, where there are great differences in levels, lifts may be needed because ramps would be either too long to fit on the platforms or too steep to negotiate. A small ramp (drawbridge) can be used to overcome small level differences between vehicle and platform. Such ramps may also be necessary to span large gaps between the platform and vehicle that trap the front casters of wheelchairs or are hazardous for users of walking aids and people with visual impairments (Kanbayashi, 1999). Although it is desirable to reduce such gaps to a safe size, larger gaps may be necessary to accommodate the lateral tolerances of rolling stock, especially if the vehicles used on one track have different widths.

Many different approaches are being used throughout the U.S. and in other countries to address the gap problem. In express bus systems with platform loading and commuter rail systems, some providers have installed hinged folding ramps or telescoping "bridge-plates" on vehicles that are manually operated by conductors or station personnel. Some bus system operators have devised compressible bumper systems that are fastened to the platform edge. Vehicles dock against the bumpers, which have some resilience and can be rolled or walked over when compressed. Other operators have equipped each station with a portable ramp or lift. In the old commuter rail stations of the SEPTA system, for example, the portable ramp approach is used to accommodate vertical and horizontal gaps (see Figure 5.9a). The ramps are located on the platforms in a locked storage system where they are readily available for conductors to use when needed. SEPTA is now including folding ramps in their new rail cars (see Figure 5.9b).

Changing Platform Levels

Raising the entire secure area of transportation terminals off the ground to the same level as vehicle floors is a good universal design strategy. It requires passengers to negotiate only two level changes—at the end and at the beginning of the trip. Moreover, this strategy benefits all riders by increasing safety and convenience and improving service response by reducing the time of loading and unloading a vehicle. Finally, it makes it easier to maintain security in the system since no one can get on vehicles unless they are on a platform. All entries to platforms can be controlled easily. Typically, platform loading is used only in rail systems but bus systems can also benefit from the use of this strategy. In Curitiba, Brazil, all express bus terminals and stops are raised off the ground and are accessible by ramp or lift as well as stairs.

In some transit systems, "mini-platforms" are used instead of full platforms to provide access to light rail systems. These above-ground stations have a small raised platform level with the vehicle floor and equipped with a ramp or lift. However, the platform only serves one car in a train to keep the platform length to a minimum. This approach, while less expensive, is not effective because if one car is full, people who cannot walk stairs have to wait for the next train. Moreover, the operator has to dock the train precisely at the platform, which, unless automated, is not easy to do.

The individual bus stations and terminals on the express bus routes in Curitiba are all fully accessible. High-floor buses and raised platforms reduce the need for negotiating level changes. The stations and terminals all have accessible entries. Once riders enter a station, they stay at the higher level and never have to use stairs or any other vertical circulation until they leave the system. All vehicles are loaded with great ease from the platform. Automated bridge-plates extend from the buses when they dock at terminals to eliminate the gap.

Safety Issues at Platforms

Avoiding falls off loading platforms is not only a major safety concern for individuals with visual impairments but is also a safety concern for all travelers. There are several methods that can be used to protect people from falling. One is the use of a gate and barrier system. This is by far the safest strategy, although it constrains the location of loading and unloading. Other safety measures include the use of warning signals and tactile warnings on floor surfaces. Warning signals provide advanced notification that vehicles are about to arrive at the platform. Visible, audible, and tactile signals should provide redundant modes of information. In Washington DC Metro stations, the platform edge is marked by a row of lights embedded in a strip of granite about 18-in wide along the platform edge (Figure 5.10). When a train is about to arrive, the entire row of lights starts flashing. However, there is no companion audible signal or announcement. Other rapid transit systems have verbal announcements but often do not have companion visual signals.

Platform edges can be marked with a contrasting color and change in texture. Many different textures have been used by transit authorities. They include rough concrete or stone, applied resilient plastic materials, and contrasting paving materials like concrete against brick. Controlled experiments have demonstrated that tactile warnings can be detectable to visually impaired travelers and are not barriers to people who have mobility impairments (Bentzen, Nolin, Easton, Desmarais, & Mitchell, 1994; Hauger, Rigby, Safewright, & McAuley, 1996). However, research has not been conducted during actual use to determine if they work effectively in actual transportation environments where there are large crowds and they may be approached at an angle rather than from a true perpendicular direction (Richmond & Steinfeld, 1999).

Many professionals and consumer advocates question the relative reliability and safety of tactile tiles compared to other approaches. The textures may have to be very obtrusive to work with a high degree of reliability. However, exaggerated textures could cause tripping hazards for pedestrians. An important consideration in utilizing tactile tiles is

Figure 5.10 Image of DC Metro platform with embedded lighting.
Source: IDeA Center

to ensure that people with visual impairments can distinguish between an edge warning and tactile guide paths. Guide paths on platforms should keep travelers away from the dangerous part of the platform (the edge), but the tactile edge itself can be perceived as a guide path putting the pedestrian in danger. Another related problem is that trains have gaps between cars that appear, to the blind traveler to be the entries of the vehicle. Thus, it is not sufficient to clarify the edge of the platform. The location of vehicle entries needs to be identified as well. Kanbayashi (1999) points out that tactile guide tiles along a platform are not very useful as a navigation aid because people and luggage create obstacles along them. He also reports that people with visual impairments still fall onto the track despite the presence of these tiles at the edge of the platform. One such accident in Washington, DC, where tactile warnings were employed, demonstrates that protecting people from falls on platforms is a problem that may not yet be solved (Weir, 2009). In this incident, which was recorded on video, a blind man walked off a station platform. He approached the platform at an angle; neither his cane nor foot touched the tactile edge before he lost his balance.

The most secure system for protecting waiting passengers at the platform edge is a physical barrier along the entire platform. In new subway stations guard rails and even complete glazed enclosures with automated sliding doors are being introduced. Not only do such barriers protect people with visual impairments, but they also protect the general population from being pushed off the platform and they prevent suicides by jumping in front of trains. Current technology supports precise docking of vehicles that enables this approach to platform safety. This strategy will not work where trains have variable lengths unless the doors are individually controlled. In older systems, where existing trains do not have computer control, only new cars can be used on the lines equipped with barriers. This may be too great a constraint for a large transit system that needs flexibility in deploying vehicles. In a new extension of an existing subway line being constructed in New York City, full height barriers were rejected for these reasons.

Problems

- Providing accessible, safe, and secure pedestrian access to stops and stations.
- Providing safe, secure, and comfortable bus shelters.
- Addressing complex information needs in transit terminals and stations for people with sensory and communications impairments.
- Questions about the effectiveness of current technologies used to provide safety and guidance information for people with vision impairments.
- Proving the effectiveness of universal design strategies for station and terminal design.

Recommendations

- Update bus shelter designs to include more innovative approaches to design and construction (e.g., franchise-bid programs).
- Compile and utilize existing reports on current practices for improving access to bus stops.
- Explore how research on tactile warning signals and RIAS, which has thus far been limited to controlled studies, can be applied to real-life settings.

Directions for Future Work

- Identifying best practices in design and maintenance of accessible and safe pedestrian access to stations including coordination with public works departments and private landowners.
- Studying the effectiveness of franchise-bid programs in providing accessible, safe, and secure bus shelters throughout a city.
- Studying the effectiveness of tactile warning signals, guide strips, and RIAS in actual transit environments and over the long term.
- Identifying and evaluating current strategies to provide dynamic information for people with sensory and communications impairments.
- Post occupancy evaluations of transit stations and terminals with universal design features.

Photo of an individual boarding a bus using a ramp.
Source: IDeA Center

6   Vehicle Design

James Lenker, Clive D'Souza, and Victor Paquet

Overview

The accessibility of public transportation services is directly impacted by vehicle design. In particular, design features associated with vehicle boarding, use, and disembarking (B/D) are aspects of the Travel Chain that can greatly impact passengers' travel experiences, and their choice of whether or not to use public transit service. With that in mind, this chapter discusses vehicle design elements associated with the interrelated steps associated with B/D, which include: negotiating a level change from grade or platform to vehicle floor at ingress, fare payment, interior circulation down a primary aisle, finding and taking a seat (or, in the case of wheeled mobility device users, securement in a designated area), exiting the bus, and navigating a level change at exit.

Transit buses can be broadly characterized based on type (e.g., high floor, low floor, articulated), location and number of entries/exits, and overall size. Buses have considerable diversity in their interior configurations in terms of occupancy (seated vs. standing capacity), seat orientation (side-facing, forward-facing, rearward-facing), seat spacing, wheelchair securement locations, and assistive features (e.g., handrails, stanchions). Each of these design choices can affect the safety, efficiency, and enjoyment of the overall transit experience.

Recent studies substantiate ongoing problems with boarding and disembarking that are experienced by transit bus riders with mobility impairments (Albertson & Falkmer, 2005; Frost, Bertocci, & Sison, 2010; Frost, Bertocci, & Smalley, 2015; National Highway Traffic Safety Administration, 1997; Nelson\Nygaard Consulting Associates, 2008). Among wheeled mobility equipment users living in areas served by public transit, 40 percent indicate that they have wheelchair or scooter access problems with public transit (LaPlante & Kaye, 2010).

Recommendations discussed in this chapter include options for minimizing vertical height differences between vehicle and environment to achieve ramp slopes no steeper than 1:8, use of electronic fare payment systems instead of cash payment systems, and increased securement area space to accommodate the dimensions of larger wheelchairs and users. Designers, transit authorities, and policy makers should also consider how the interaction of components in the bus and the external environment impact passenger B/D performance when considering public transportation accessibility.

New ramp deployment and storage systems are needed to address tradeoffs between ramp length and storage requirements. New wheelchair securement systems and electronic systems are also needed that simplify the securement process for passengers and drivers. The field would benefit from assessment tools that better characterize the usability issues presented by different interior bus configurations, and a database of B/D usability for different interior designs and environmental contexts. Additional research is also needed to explore the interaction of bus design, environmental conditions, and passenger abilities on B/D performance.

While this chapter is organized by specific design features commonly found in local public transit buses, it is important to recognize that everyday use of these vehicle design features is interrelated. For example, a low-floor bus may eliminate the need to use stairs during B/D, but may reduce the space available for maneuvering toward the front of the bus where the ramp meets the bus floor and between the bus wheel wells. Manual fare payment systems may also require floor space that can further compound the design of the entry.

Change in Level for Ingress and Egress

There are four options to facilitate the change in level required to enter and exit a bus: stairs, electromechanical lifts, ramps, and platform loading. Traditional high-floor transit bus designs require ingress/egress via stairs for ambulating riders and electromechanical lifts for wheelchair users. The lifts are usually installed at the rear of the bus to allow use of stairs at the front entry, though there are some lifts that are integrated with stairs or stored under stairs in the front. Buses used for inter-city travel are often equipped with lifts having different vertical ranges, since these buses have storage compartments under the passenger compartment.

High-floor buses do have their advantages for specific applications, namely higher passenger capacity and a higher underside clearance. They can be used effectively in BRT systems and other applications with platform boarding. In these applications, the key accessibility issue is the horizontal platform gap. There are known solutions to the horizontal gap problem, including soft edge bumpers that allow the bus to actually touch the edge of a platform, small drawbridges attached to the bus or the platform, and robotic docking systems that provide accuracy sufficient to reduce the gap to an acceptable amount (see Chapter 5).

Historically, step entrances in transit buses presented a barrier to boarding and disembarking for wheeled mobility users. Electromechanical lifts were initially used to address this accessibility barrier; however, first-generation lifts were considered unsatisfactory because they were prone to breakdown, required bus driver assistance, created long loading and unloading delay, and were not helpful for ambulation aid users. Although reliability has improved, the other limitations remain. The emergence of low-floor bus designs in the late 1980s lowered the floor height by approximately 60–70 cm (Blennemann, 1991), reducing physical demands and tripping risks (Rutenberg, 1995; Schneider & Brechbuhl, 1991). Many low-floor buses also "kneel" at stops, further reducing the initial step height by 7–10 cm, resulting in an entry step of 24–28 cm above grade. The overall reduction in ground-to-bus floor height has made it feasible to replace lifts with access ramps (Rutenberg, 1995).

The prevalence of high-floor buses for local public transit is thus declining and will continue to decline with the proliferation of low-floor designs that offer more efficient entry and egress with access ramps (see Figure 6.1).

Compared to wheelchair lifts, access ramps have a simpler design that is less prone to breakdown and requires less maintenance (Blennemann, 1991; Rutenberg, 1995; Schneider & Brechbuhl, 1991). Ramps enable wheeled mobility users to board vehicles more discreetly and in less time (Blennemann, 1991; Rutenberg, 1995). For drivers, ramps are simpler to deploy and do not require them to leave their seat (Rutenberg, 1995; Schneider and Brechbuhl, 1991) although some manual chair users require assistance. Ramps can also be used by ambulation aid users, parents pushing strollers, and riders with rolling suitcases or shopping carts, allowing a greater percentage of passengers to enter and exit the bus with reduced effort and assistance (Rutenberg, 1995; Schneider & Brechbuhl, 1991).

In the APT field, the slope of access ramps has been a contentious issue. Steeper slopes (a) increase the level of physical effort associated with ambulating or propelling "uphill" during ingress, (b) induce eccentric workload that is associated with controlling locomotion speed during ramp descent, and (c) accentuate challenges to standing balance (for those who ambulate) and sitting balance (for wheelchair users) that are caused by uphill and downhill locomotion on a sloped surface.

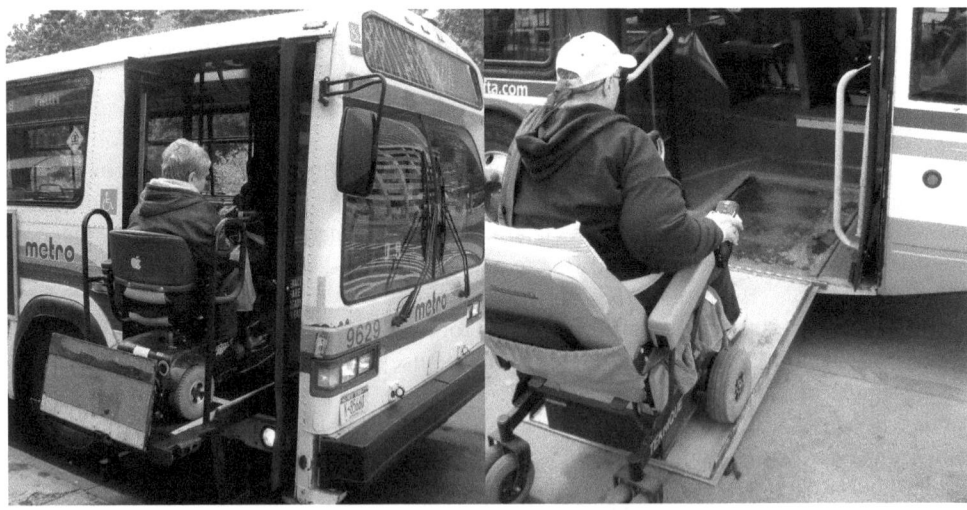

Figure 6.1 (a) Lifts used in traditional high-floor buses to allow passengers who cannot board or exit via stairs, but are costly to maintain and increase bus dwell times at local stops, (b) ramps used in low-floor buses are less expensive to maintain and generally allow more efficient boarding and exiting of the bus.

Source: IDeA Center

Access ramps have several other drawbacks. Drivers must alert those waiting outside that ramp deployment is imminent. In the U.S., the majority of ramps are mounted at the front of the bus where the space is constrained by the driver's seating area, fare payment system, and the wheel wells. Longer ramps that result in reduced slopes may reduce the floor space available for fare payment and maneuvering from the ramp to the passenger seating area. Ramps require substantial horizontal space when deployed, which creates a design challenge for manufacturers to achieve compact ramp storage in the limited space of the bus.

Ramp design issues create a tension for policy makers, who must attempt to balance the accessibility needs of mobility aid users with the ramp design challenges facing manufacturers. People with mobility impairment prefer gentler slopes; however, ramps with gentler slopes are more difficult to design. To achieve gentler slopes, ramp and bus manufacturers must create ramps of greater length that can be folded and stowed in a space that is inherently constrained by available floor space at the bus entrance and the powertrain and vehicle suspension components that are underneath the floor.

Previous Americans with Disabilities Act Accessibility Guidelines (ADAAG) for Transportation Vehicles stipulated that ramp slope could vary from 1:4 to 1:12, depending on the overall rise (U.S. Access Board and Department of Transportation, 1998). In 2010, the U.S. Access Board started a rule-making activity for public transportation that proposed a maximum allowable slope of 1:6, regardless of deployment conditions. Subsequently, advocacy groups and professionals expressed concerns that the proposed 1:6 ramp slope criterion would pose safety and usability problems for some users (U.S. Access Board and Department of Transportation, 2007b). Many industry representatives commented that it would be too restrictive to be feasible, given design constraints (U.S. Access Board, 2007b) and inconclusive research literature. The Access Board issued a final rule in late 2016 that adopted the 1:6 maximum slope, compliance with which will become mandatory when the U.S. Department of Transportation (DOT) next updates the ADA.

The associate research literature evaluating the effects of access ramp slope on B/D has been hampered by inconsistent methodological elements (e.g., sample populations that are narrowly defined or vaguely described; ramp slopes and lengths that do not reflect the ranges relevant to transit buses; measures that yield a relatively narrow perspective on usability evaluation) across studies (Nelson\Nygaard Consulting Associates, 2008). Frost & Bertocci (2010) evaluated 115 adverse incidents involving wheeled mobility devices on large accessible transit buses over a six-year period in Louisville, KY, and found that 42.6 percent (*n* = 49) were associated with ingress/egress. Among these, 12 of 49 involved the wheeled mobility device tipping forward or rearward while ascending or descending the access ramp, prompting the authors to conclude that, "… research is needed to examine the adequacy of existing federal legislation and guidelines for accessible ramps used in public transportation (ibid, p. 236)." A subsequent study of boarding and alighting (Frost et al., 2015) found that five percent of wheeled mobility device users experience a ramp-related incident when accessing public transit buses and that these incidents were five times more likely when the ramp slope exceeded 9.5° (1:6).

A recent laboratory study of manual wheelchair users, powered wheelchair users, and individuals with visual impairments who used a cane or service animal was completed to address these limitations (Lenker et al., 2016). Results showed that the steeper ramps allowed by current standards reduced boarding and disembarking performance, and that performance is impacted differentially across passenger groups. The results suggested that the 1:4 and, in some cases the 1:6 slope, were too steep for safe, unassisted, and/or efficient boarding. Manual wheelchair users perceived greater exertion and greater difficulty, and the differences were more pronounced as slope increased compared to other passenger groups. Power wheelchair users and people with vision impairments expressed concerns about descending ramps with steeper slopes. The results suggest slopes of 1:8 may result in improved boarding and disembarking performance among these passenger groups compared to slopes of 1:4 or 1:6. It is important to note that the data were collected under two idealized conditions: (a) the test apparatus was located in an indoor laboratory setting that did not reflect adverse outdoor weather conditions (e.g., snow, ice, rain, wind) that would presumably degrade performance across slope conditions under real-world weather conditions; (b) the width of the test ramp was 102 cm, a wider-than-typical dimension chosen to eliminate the potentially confounding effects of narrower ramp widths on the ability to navigate different grades. The data from this study thus suggest a baseline of best-case performance that can be a useful basis for comparison with future data captured in real-world environments.

The slope of deployed access ramps is not purely a design burden for bus manufacturers. A variety of environmental design factors also contribute to the ramp slopes achievable in everyday situations (e.g., availability of raised platforms, accessibility of bus stops and sidewalks leading to bus stop areas, illegally parked cars that block sidewalk deployment of ramps at bus stop areas, and accumulations of snow at bus stop areas during winter months). Regulatory authorities, policy makers, and planners should thus consider additional strategies for minimizing ramp slopes deployed in everyday operating conditions. Currently, ramp manufacturers have created designs that achieve 1:8 slopes when deployed to sidewalks or raised pads. To support this technology, bus stops could be designed to make it easier for bus drivers to stop near the curb, which makes it possible to achieve a consistent 1:8 deployment slope to the sidewalk or pad rather than the road. For example, one strategy would be to create bulb-outs at bus stops. Bulb-outs effectively bring the sidewalk closer to the path of vehicle travel and would carry the additional advantage of calming traffic and creating a safer pedestrian environment. Likewise, more attention can be given to reducing illegal parking in loading areas. In suburban areas, sidewalk construction is an important link in the overall Travel Chain. In rural areas, devising less expensive means to construct bus stop pads is an important priority. These strategies emphasize the importance of coordination among transit agencies, public works departments, public safety authorities, and advocates for improved walkability (Figure 6.2).

Figure 6.2 Steeper ramp slopes such as those encountered when a ramp is deployed to the roadside present more usability problems than shallower ramp slopes that result when a ramp is instead deployed to sidewalk or bulb out.

Source: IDeA Center

Floor Plan Configuration

Previous studies evaluating the accessibility of transit vehicle interiors have generally taken one of two approaches: (a) exploratory field studies that queried transit riders on problems experienced during boarding and disembarking (e.g., Wretstrand, Stahl, & Petzall, 2008); or (b) lab testing of interior component systems in isolation (e.g., Petzall, 1993; Daamen et al., 2008). Field studies are advantageous because they utilize real-world environmental conditions (e.g., moving vehicles and passenger interaction) that provide insights regarding everyday hazards and human performance in B/D. However, these conditions introduce uncontrolled factors that complicate data interpretation, and are limited to observations of only individuals who are capable of accessing and successfully using public transportation. In addition, field studies are unavoidably constrained by the vehicles being studied, i.e., field researchers are unable to manipulate design features in a manner that allows direct comparisons. In contrast, laboratory studies afford tightly controlled component-level evaluation of features in relative isolation.

Full-scale laboratory simulations can capitalize on the advantages of both field studies of bus B/D and laboratory research that evaluates individual design features. One such study evaluated the usability of interior bus floor plan configurations, demonstrating large differences on B/D performance across floor plans, crowding conditions, and passenger mobility aid user groups, to varying degrees (D'Souza, Paquet, Lenker, Steinfeld, & Bareria, 2012; D'Souza et al., 2017; Bareria, D'Souza, Lenker, Paquet, & Steinfeld, 2012). The study used a full-scale mock-up of a transit vehicle and mannequins to support a realistic but controlled evaluation of multiple interior conditions (e.g., seating layout, doorway location, passenger density, and design features) on the overall passenger B/D experience. Among three different interior bus configurations (Figure 6.3), the greatest efficiency and safety for wheeled mobility device users was found for a Mid Entry-Mid Exit configuration that had an electronic on-board fare payment, rear-bus entrance doorways, and adjacent device securement areas. A Mid Entry-Forward Exit (two accessible entries/exits) resulted in high efficiency in the low passenger crowding condition, but lower efficiencies and poorer safety for the high passenger crowding condition. A Forward Entry-Forward Exit configuration, the most common configuration used in the U.S., resulted in the poorest efficiency (D'Souza et al., 2017).

The results suggest that more can be done to enhance the B/D efficiency of wheeled mobility device users in the U.S. However, use of Rear Entry-Rear Exit and/or Rear Entry-Front Exit configurations requires design changes to the built environment (i.e., bus stops) to accommodate the change in the relative positioning of the bus to the local stop. It would also require changes to the design and location of fare payment systems, which are often located in the front of transit buses used in the U.S. Further implications of vehicle design are described in other sections below.

The impact of vehicle interior design on B/D performance is directly relevant to transit agencies, most of whom produce their own specifications for manufacturers when ordering new vehicles or retrofitting existing vehicle interiors. These specifications are based on the practical experience and specific needs of transit agencies combined with the particular floor plans available from the manufacturer that also comply with federal accessibility standards (e.g., U.S. DOT, 2007) and industry guidelines in bus procurement (e.g., APTA, 2009). Individual transit agency's specifications are shared with other transit agencies but are not generally distributed. Thus, there is no coordinated effort to evaluate combinations of features of the buses or local stops for usability, nor is there a centralized repository where such information can be obtained.

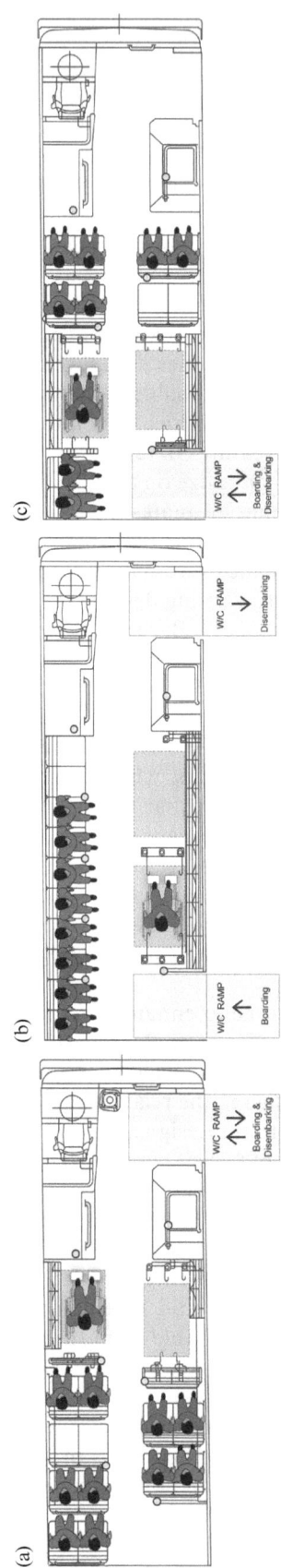

Figure 6.3 Plan views of the front half of three interior bus configurations, along with mannequin placement for simulated conditions of low and high passenger loading (D'Souza et al., 2017). (a) Front entry-forward exit configuration with forward-facing seats, (b) rear entry-forward exit with side-facing seats, (c) rear entry-rear exit with forward-facing seats.

Source: IDeA Center

Entry/Egress Location

The location of the entry and egress doorway influences the choice of seating configuration, location of on-board fare payment systems, and positioning of the wheelchair securement area (D'Souza et al., 2012; D'Souza et al., 2017). The presence of adequate support features (e.g., handrails and grab-bars) also affects entry and egress, particularly for older adults and those with mobility impairments (D'Souza et al., 2012; Petzall, 1993). The horizontal and vertical gap between the platform and vehicle floor, door width, and payment method also impact the problems encountered by people with disabilities during entry and egress (Daamen et al., 2008) and overall dwell time (Fernández et al., 2010).

Accessible boarding at mid-vehicle entries can eliminate many of the maneuvering problems caused by protruding front wheel wells at the forward area of low-floor buses. Rear or mid boarding is the favored approach in Europe. However, mid-boarding has not been widely implemented in the U.S. due to concerns of remote or unsupervised ramp/lift deployment and fare payment validation. These options are largely limited to transit systems that have platform boarding with either pre-paid fare or no-fare (e.g., BRT, airport, and amusement park shuttles).

In the future, combinations of new technologies may create opportunities for accessible boarding at multiple entrance locations, which would increase passenger throughput and reduce dwell times. For example, electronic fare payment combined with automatic passenger counting (APC) systems that offer directional counting of passengers and computer vision techniques to automatically recognize fare payment gestures can provide real-time validation and detect fraud. On-board cameras and ramps equipped with sensors that detect surrounding movement and distance combined with safety interlocks and alert warnings could increase safety and driver situation awareness during remote ramp deployment. Such advances could allow passengers to board and exit from front and mid-vehicle entry-exit locations (Figure 6.4).

Fare Payment

On-board fare payment systems can be classified into two groups: *cash-based* and *electronic*. For both groups, payment machines in the U.S. are typically located at the front of the bus to ensure fare payment validation and security.

Cash-based payment systems are not well integrated into the interior bus design, often encroaching into the aisle and affecting available clearances near the front doorway. This limits maneuvering space for wheelchair users when turning into the aisle during boarding and also from aligning with the access ramp during disembarking increasing the risk of tipping or rolling off the ramp.

Electronic fare payment systems, which are increasingly popular with transit agencies, generally have two forms: (a) *contact-based* cards ("swipe-cards") which are small plastic cards with a magnetic stripe that are swiped through a slot on the fare-box; and, (b) *contactless* proximity cards, which include an embedded radio frequency identification (RFID) chip that is recognized when held close to or waved near a secure card reader. Compared to cash-based payment systems, electronic fare payment can greatly reduce the physical (e.g., reach distance and manual dexterity), visual, cognitive (counting change, remembering to bring change), and emotional (time pressure) demands on passengers (Figure 6.5). As electronic fare payment systems become more compact, it will be important to provide adequate tactile and auditory feedback (e.g., handrails, earcons) to assist blind and visually impaired users in locating and using the card reader (Bareria et al., 2012).

Technologies that further integrate fare payment hardware with the vehicle's physical environment can reduce the challenges of on-board fare payment, increase available

Figure 6.4 Ramp deployment for an entry-exit located near the middle of the bus allows passengers more room when maneuvering inside the bus.
Source: IDeA Center

floor space, and improve overall usability for all passengers, including those with disabilities. For example, these benefits can be achieved by mounting fare machines and card readers on the front dashboard or embedding them in the dashboard panel near the entry handrail. This approach requires creation of modular payment systems with standardized hardware interfaces to facilitate payment.

Agencies are also moving toward off-board payment systems located at transit stops (Transportation Research Board, 2012). In this scenario, passengers buy their ticket at machines installed at the stop and then show proof-of-payment to the driver upon boarding. This approach reduces boarding time and eliminates the need for an on-board fare machine.

Location and Size of Wheelchair Securement Spaces

Wheelchair securement spaces are often situated close to accessible boarding locations to minimize problems with interior circulation and wheeled mobility maneuvering clearances. Most securement spaces also contain seats for ambulatory passengers that can fold-up when a wheeled mobility user requires the space. Wheelchair securement spaces can differ in terms of location (e.g., front entry and rear entry) and orientation (e.g., front-facing vs. rear-facing securement), which together result in vastly different maneuvering clearances for wheeled mobility device users. This also imposes constraints on interior circulation patterns of ambulatory users, priority seating for the elderly, and the space available for other large devices such as walkers, Segways, strollers, and carts.

Figure 6.5 Electronic fare reader reduces fare payment time but when combined with a cash payment option requires a safe below, thus reducing clear floor space at the bus entry.
Source: IDeA Center

Current accessibility standards prescribe minimum dimensions of 76 cm × 122 cm (30 in. × 48 in.) for securement spaces based on mobility device size. The securement space provided on buses rarely exceeds these minimum requirements. Research on the anthropometry of wheeled mobility devices indicates that these dimensions are inadequate for accommodating many users of wheeled mobility devices, especially users of power chairs and scooters (Steinfeld, D'Souza, & Maisel, 2010; D'Souza et al., 2010). The mismatch between equipment dimensions and available space is becoming more prevalent. Increasing numbers of wheelchair users have complex needs that necessitate advanced postural support, elevating leg-rests, and longer and/or wider wheelbases, all of which result in a larger occupied footprint. The growing prevalence of power chairs and scooters is expected to further change the mobility device user demographics in the coming years (Cooper, Cooper, & Boninger, 2008). Securement spaces that do not accommodate larger wheelchairs may preclude some passengers from using a bus, make it difficult for others to position their wheelchair within the space, and increase dwell times at local stops.

New guidelines are needed to accommodate large mobility devices on transit vehicles. Such guidelines should be sensitive to local contexts that may have differences in ridership characteristics (volume and composition) and vehicle design (size and diversity of the fleet). Hence, transit agencies will require guidance to incorporate a "best practice" strategy that meets their unique needs. Standards development work is underway on modifying the clear floor space requirements in the current consensus accessibility standards to reflect the prevalence of larger occupied wheeled mobility devices (ICC, 2009).

Wheelchair Securement Systems

In many countries, buses are required to have securement space for only one wheeled mobility device. In contrast, the U.S. DOT's ADA regulations requires all ADA-compliant buses to have two wheelchair securement areas, each having the ability to secure the wheelchair and a seat belt and shoulder harness to secure the wheelchair user. Transit authorities determine whether or not to require use of these securement features by all passengers who use wheelchairs. Given the option, many wheeled mobility users prefer not to be secured.

Previous research and development has led to safer securement systems and adoption of voluntary industry standards for wheeled mobility devices used in transportation (Frost et al., 2012; Karg, Buning, Bertocci, Furhman, Hobson, & Manary, 2009; RERC-WTS, 2012b; Schneider, Manary, Hobson, & Bertocci, 2008). The four-point belt-type tie-down system is the most common method of wheeled mobility device securement systems on transit buses. This is partly due to standards requirements and its flexibility. A key requirement for its proper deployment is the identification of appropriate attachment points on the wheelchairs – a task often charged to the bus driver. Wheelchairs complying with ANSI/RESNA WC19 *Wheelchairs Used as Seats in Motor Vehicles* (ANSI/RESNA, 2000) are equipped with clearly designated attachment points on the chassis of the wheelchair. However, the practice of wheelchair securement still remains low at around 39 percent (Buning, Getchell, Bertocci, & Fitzgerald, 2007).

Frost, Bertocci, and Salipur (2013) summarized several studies that reported usability limitations of tie-down systems, including discomfort, unwillingness, lack of knowledge among drivers regarding safe and accurate securement, and the profusion of wheeled mobility device designs and accessories that hamper proper securement. Tie-down systems often require significant driver assistance and involvement. For drivers, the attachment of tie-downs requires them to invade the personal space of passengers and adopt awkward postures that pose an injury risk. More information is needed regarding the challenges of securement from the driver's perspective, as well as passenger concerns about safety, esthetics, and comfort (Buning et al., 2007; Steinfeld, Grimble, et al., 2012). It is also important to explore the reasons why the WC19 standards have not been fully adopted by mobility device manufacturers and consumers, which may include the impact of the standard on usability, convenience, aesthetics, and costs of wheelchairs, as well as knowledge of the standard among clinicians and third-party reimbursement rates for WC19 devices.

Improvements in wheelchair securement practice are also possible through novel implementation strategies. For example, some agencies have instituted a "tether strap program," which involves a one-time appointment wherein trained professionals attach markings or tether straps on an individual's wheelchair to unambiguously designate the attachment point for securement, thus expediting on-board securement (Cross, 2011). Providing transit agencies with required technical assistance and personnel to support such outreach efforts is equally important.

Ongoing usability evaluation of newly emerging securement technologies is also necessary. These include securement systems designed for independent use that could result in more efficient B/D such as with a docking interface, e.g., the Universal Design Interface Geometry (UDIG) adaptor (Hobson & van Roosmalen, 2007; van Roosmalen, Karg, Hobson, Turkovich, & Porach, 2011) or with a movable aisle-side arm for containing the device, e.g., the Bus-Buddy (LINC Design LLC, 2012). Rear-facing wheelchair passenger stations (RF-WPS) are

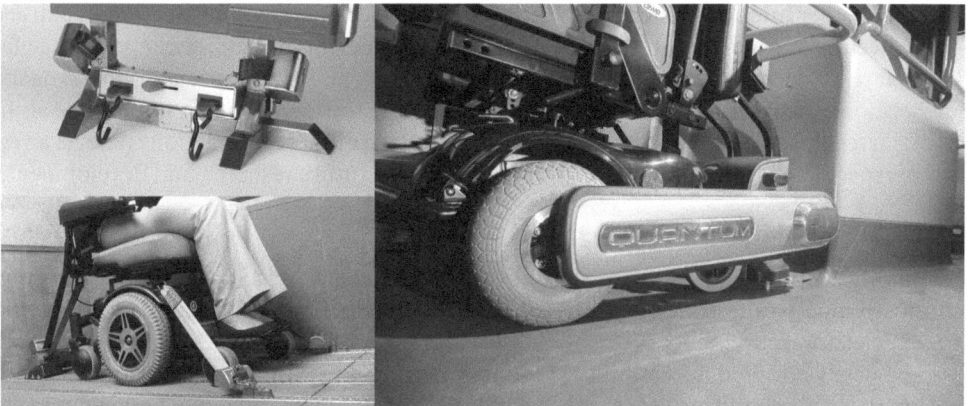

Figure 6.6 Traditional and alternative wheelchair securement systems.
Source: Photos courtesy of Q'Straint

also a viable solution in low-g environments (e.g., large urban transit vehicles) and are widely accepted in Canada and Europe (Rutenberg & Hemily, 2003). RF-WPS can be equipped with passive (i.e., a rear head-board with a fixed stanchion on the aisle-side) or active securement such as with a pneumatically-activated aisle-side arm and wall-side contact plate (van Roosmalen et al., 2011; Turkovich, van Roosmalen, Hobson, & Porach, 2011). An active securement RF-WPS system designed for independent use (via push-button activated aisle and curb-side movable arms) is now commercially available (Q'Straint Inc., 2012).

One such study (Perez, Kocher, Nemade, Paquet, & Lenker, 2016) compared the usability of three securement systems: a conventional forward-facing four-point securement system (4-Point), a three-point forward-facing compartment system (Q-Pod), and an automated rear-facing securement system (Quantum). Data were collected from three user groups (manual wheelchairs, power wheelchairs, and scooters) in a lab setting utilizing a full-scale mock-up of a low-floor transit bus. The results indicated that the Quantum was easier to use, faster to use, and required less assistance to use than the conventional-facing and Q-Pod systems (Figure 6.6). Although the study did not field-test the securement systems under everyday conditions, the findings nonetheless suggest that emerging wheelchair securement technologies may provide viable alternatives that increase acceptance and use of securement systems by drivers and passengers. More research and development is needed on wheelchair securement systems that increase independence for passengers using mobility devices and decrease dwell time at local stops.

Conclusion

Accessibility of transit vehicles has been achieved through federal standards and guidelines (e.g., U.S. DOT, 2007) that prescribe minimum requirements for essential equipment and spatial clearances (e.g., doors, ramps and lifts, aisle widths, wheelchair securement systems, and communication systems). This has led to great progress toward more accessible transit vehicles, but there is room for improvement. Advances in ramp design, bus configurations, including entry and exit locations, wheelchair securement systems, and local stops may accommodate the needs of even more passengers with functional limitations. It should also be recognized that these components are not used in isolation—they are, in fact, used in rapid succession by passengers engaged in vehicle boarding and disembarking (B/D).

In everyday situations, the interactions among equipment arrangement, design features, space, and other interior elements (e.g., size of the interior, number and location of entries/exits, wheel well covers, number and type of seats, placement of seats, package storage facilities, fare payment systems, handholds) impact the overall B/D experience of passengers. Furthermore, the interaction between the bus and bus stop environment impact the abilities of passengers to board and exit buses. Consideration of these interactions reflects a *systems* perspective that should be given more attention in future research and development.

Problems

- Bus dwell time can increase significantly when mobility aid users are boarding, mostly due to the time required for ramp deployment, ascent, ramp storage, and wheelchair securement.
- The newly adopted accessibility standards require a maximum access ramp slope of 1:6 under all conditions (U.S. Access Board, 2016). Meanwhile, bus and ramp manufactures are not in agreement about the feasibility of providing ramps that ensure a maximum 1:6 slope regardless of the environmental condition. Ramp slopes of 1:6 may provide significant difficulty for some passengers, particularly those using wheeled mobility devices.
- Access ramps must be stored within the floor of the transit bus. For bus manufacturers, this creates substantial design challenges that are accentuated by the greater ramp lengths that are necessary to achieve shallower ramp slopes.
- Wheeled mobility devices, in particular powered mobility devices, are increasing in clearance width and effective length (Steinfeld et al., 2010), which amplifies challenges associated with B/D. Larger individuals and their wheelchairs sometimes need assistance on ramps and exceed the size and weight capacity of ADA-compliant lifts (600 lbs). In response, transit providers are implementing exclusion policies based on weight and size (NCD, 2005), while some are requiring higher capacity equipment (APTA, 2007a). There is also a general inconsistency in the operational policies and practices of different transit agencies for accommodating and prioritizing available space for different large devices (Goldman & Murray, 2011).
- Most ramp-equipped low-floor buses in the U.S. only permit the front doorway for accessible entry and exit; however, other accessible entry/exit configurations (e.g., mid-bus) are possible and would provide opportunities for more efficient passenger boarding and disembarking. These, however, might also require substantial changes to the design of local stops.
- The ability to identify suitable and/or improved bus configurations and design features based on rider characteristics is hampered by the lack of a centralized information resource and systematic study of different designs and environmental contexts.
- Floor-mounted fare machines occupy valuable floor space and also impede approach and usability of wheeled mobility users and visually impaired users in particular.
- Despite the safety implications, the practice of wheelchair securement remains infrequent and is still perceived as cumbersome by passengers and drivers. New securement designs have promising features but require further usability and safety evaluation before widespread implementation.

Recommendations

- Transit authorities should identify mechanisms for accommodating those with mobility impairments without substantially compromising overall system efficiency and reliability. The incentive is high for local transit authorities, given that the next best alternative mode (i.e., paratransit) is very costly.

- It is inherently difficult to achieve maximum access ramp slopes of 1:8 in all roadway surface conditions with current technologies. Regulatory authorities, policy makers, and planners should thus place greater attention on how to achieve a ramp slope of no greater than 1:8 through all means available (e.g., environmental modifications, including redesign of bus stops) rather than the ramp alone.
- The accessible public transportation field needs to determine what level of mobility aid accommodation is realistic and practical with regard to contemporary vehicle technology with due consideration of the technological constraints that exist within the transit environment, and the increasing size of occupied wheeled mobility devices. This can be used to guide short-term and long-term accessibility design solutions.
- Designers, transit authorities, and policy makers should consider how the interaction of components in the bus and the external environment impact passenger B/D performance when considering public transportation accessibility.

Directions for Future Work

- New ramp deployment and storage systems are needed to successfully address tradeoffs between the need for ramp length (which must be longer for shallower slopes), ramp storage (which is complicated by longer ramps), and the safety risks for mobility aid users that are posed by abrupt level changes between the upper section of access ramps and the interior circulation aisle.
- Assessment tools are needed to audit new and existing vehicles for compliance with accessibility standards and guidelines given the wide diversity in vehicle interior designs (Cannon, 2011). Such tools could include standardized protocols for conducting and reporting accessibility audits of transit vehicles (Steinfeld, Grimble, et al., 2012). Developing these tools will require collaboration by manufacturers, transit authorities, mobility aid users, researchers, and policy makers.
- A database of research evidence that support efficient B/D in a variety of environmental and social contexts would be a great benefit to designers, policy makers, and transit authorities. A central repository would enable transit agencies to inform their vehicle procurement and/or retrofitting decisions with research. This could take the form of a database created from standardized usability assessments of different municipal environments (e.g., rural, suburban, urban) and interior design configurations. Such a database could allow an evaluation of the tradeoffs in vehicle design that arise from multiple and sometimes competing objectives (e.g., improved accessibility, minimizing dwell time, maximizing occupancy, minimizing costs, rider comfort) combined with differences in design needs based on ridership characteristics and type of service, e.g., conventional bus service versus BRT, on bus fare payment versus payment at the stop.
- There are a variety of research areas that require further exploration. The interactive effects of bus design and physical environment conditions (e.g., weather, topography), and individual abilities have not been fully explored. Refinement and field-testing of access ramps, electronic payment systems, and wheelchair securement systems are needed to ensure safety and usability, optimize passenger independence and comfort, and reduce dependence on paratransit.

Source: Photo courtesy of The London Taxi Cab Company

7   Demand Responsive Transportation

Edward Steinfeld and Aaron Steinfeld

Overview

The term "demand responsive transportation" (DRT) refers to modes of transport that adjust the supply of services to the demand at any one time, including variations in schedule and/or variations in trip route. DRT is an important, but often overlooked, component of public transportation that provides unique opportunities and challenges for accessibility. DRT can include paratransit, flex-route service, taxis, shuttle buses, and several other types of service. The success of transportation network companies (TNCs), like Uber and Lyft and other vehicle sharing services, is rapidly changing the DRT scene, creating both opportunities and challenges for accessible transportation.

Defining Demand Responsive

There are many definitions of demand responsive service used in the accessible transportation literature. Some sources use DRT as a synonym for special transport services

(STSs), or paratransit, as it is called in North America; some exclude STS from their definition, focusing on services available to a wider rider base; some include taxis; some exclude taxis. There are many other even more subtle variations in definition. DRT is a common form of transportation in places where traditional public transportation is not available (Enoch, Potter, Parkhurst, & Smith, 2004). It should also be noted that, although most authors focus on DRT provided through buses, vans, and taxis, it is also a form of transportation provided in the informal sector with many different types of vehicles for hire, including personal automobiles, motorcycles, rickshaws, and boats.

A report on a study of DRT in Great Britain used this four-part definition of DRT: (1) available to the general public; (2) provided by low-capacity road vehicles such as buses, vans, or taxis; (3) responds to changes in demand by either altering its route and/ or its timetable; and, (4) fare is charged on a per passenger basis (Davison, Enoch, Ryley, Quddus, & Wang, 2014). But this definition excludes shared ride services that charge by the trip rather than the person and also services where no fare is charged at all like courtesy shuttle buses. The U.S. DOT defines DRT as "a transportation system where passenger trips are generated by calls from passengers or their agents to the transit operator, who then dispatches a vehicle to pick the passengers up and transport them to their destinations" (U.S. DOT, 1988). This definition is broad but it still excludes vehicles that are hailed on the street, e.g., taxis, and services that may have a fixed route and schedule but deviate from the route and schedule as needed.

Mobile phones and crowdsourcing are disruptive technologies that are changing the nature of DRT. These two technologies have enabled the rapid development of vehicle sharing services. There are major controversies about whether these new forms of public transportation are covered by existing accessibility regulations. Are they a public service, which clearly is regulated by laws such as the ADA Act, or, are they a private transaction between two individuals not covered by the laws? Litigation and consumer advocacy are addressing this question. Whatever the courts and government decide, the current definitions of DRT in the literature overlook the growing popularity of such services, including both TNCs, like Uber and Lyft in which a private car is shared by the owner who is the driver, "time sharing" services like Zipcar, and driverless vehicles which are on the horizon.

We will use the following broad definition of DRT that avoids the exclusions of the examples above and also addresses new technologies: DRT is a transportation service available to the public that adjusts to variation in demand by alterations of schedule, route, or both. This definition includes STS provided exclusively to qualified individuals with disabilities or other categories of riders, flex-route service, courtesy shuttles of all sizes, taxicabs, and shared vehicles. It is a service mode that can be delivered with any vehicle type and size.

Special Transport Service

STS, or paratransit, is the most common form of DRT offered by public transit agencies. It is designed to supplement fixed route transit service to serve people who do not have the ability to use those services because they either cannot reach a station or stop or do not have the physical, sensory, or cognitive ability to use regular service independently. STS may also be provided where there is no public transit service at all (e.g., in a rural community). Provision of STS provided by public transit agencies is regulated by government and thus the policies governing such service will vary from country to country and from local jurisdiction to another. STS is usually provided using small vehicles that can carry from 4 to 12 riders at a time using vans or small buses equipped with lifts or ramps. The most important policies affecting the accessibility of service provided by transit agencies are related to service availability, eligibility, and service quality.

In the U.S., transit agencies are required to provide STS wherever they provide fixed route service and cannot charge significantly more for rides than the fees for fixed route service. Local transit agencies are only obliged to provide STS within 3/4 mile of transit stops. If a locality has an extensive public transportation system like New York City or Washington, DC, STS is not limited. However, in rural or suburban localities or cities with very limited transit systems, lack of service is the most significant barrier to mobility for all public transportation users. Since STS does not have to be provided in places not served by fixed route service, riders with disabilities may not have access to any public transportation options.

Transit agencies face increasing financial burdens due to STS operations as the populations they serve age and more and more riders cannot use fixed route service due to accessibility and usability barriers. For example, Washington DC's transit system estimates that their cost to provide paratransit will rise from $121 million to $171 million in a decade (Lazlo, 2016). Eligibility requirements ensure that only people who cannot use the regular transportation services are using STS. Each individual who applies for STS is evaluated although the criteria used vary. While the rider's ability plays a role in determining eligibility, so do the challenges of using the fixed route system (e.g., accessibility of vehicles, usability of fare payment, usability of schedules, maps, and websites, and the accessibility, safety, and security of the pedestrian route to the stop or station). Transit agencies can reduce such challenges significantly by using several different strategies: (1) training programs for riders, (2) increasing the number of routes and stops to reduce the distance to bus stops, (3) providing flex-route service that departs from a route to pick up riders closer to their home or at their door, (4) improving information technology like more usable websites and online reservation systems, and (5) adopting more accessible vehicles and stop infrastructure for fixed route service (e.g., low-floor buses with ramps, raised platform boarding for BRT service, etc.). Although the emphasis in the literature is on accessibility for people with mobility impairments, it is important to note that people with cognitive limitations form a major part of the STS-dependent ridership, thus investing in making the system easier to understand and reducing complexity of rides (e.g., simplifying payment, transfers, and intermodal connections, is an important strategy to address the needs of this group. Note that all these strategies benefit all users of the system).

The U.S. National Council on Disability (2005) described four common problems with STS in the U.S.: (1) vehicles do not arrive when scheduled or not at all, (2) telephone reservation systems are inadequate and present a burden for riders, (3) service providers are not responsive to complaints, and (4) inadequate information resources are provided to help riders. In particular, most systems do not have an online reservation system that allows riders to make changes independently when situations change. These problems can create significant quality of life issues for people with disabilities, including loss of employment, missing critical health services, and health risks due to overly long trips and prolonged exposure to extreme weather conditions (Frieden, 2005).

The reservation system and "no show" policies used by STS agencies is a particular source of problems for riders. Most American STS service requires making reservations at least 24 hours in advance, sometimes more. If there is a high chance that there will not be a seat available, many riders make extra reservations when they are unsure of their plans, resulting in a high volume of no shows and late cancellations. Transit agencies are allowed to suspend riders from service after a pattern of "no shows." But riders complain that when rides are late and they give up waiting and return to their homes or find alternative rides, they are unfairly blamed for a "no show." This also happens when vehicles show up early, before the rider is ready and the driver does not wait until the reservation time before leaving. There may be no system in place for documenting late service other than the driver's word. STS systems are therefore not really demand responsive in the true sense of the term, although they have the potential to be so.

Most of the problems above can be solved by better information technology solutions. Equipping vehicles or drivers with GPS systems and providing mobile phone apps that track vehicles and allow riders to communicate in real time with drivers would help to improve service quality. Such systems could also include a feature that would allow riders to rate drivers and other aspects of the service, making the providers more accountable. In addition, they could automatically document late arrivals. STS providers in many locations do not communicate with their riders to notify them about schedule delays, or, have no means to allow riders to notify drivers of last minute problems or emergencies, like exposure to very cold temperatures while waiting for a ride that is late (Frieden, 2005). Some STS scheduling systems allow riders to receive an automated phone call when the vehicle is five minutes away or less (Disability Rights Education & Defense Fund, 2010). But, it remains unclear how often this service feature, which can ameliorate problems with on-time performance, is used by service providers in the U.S.

Taxis, Hired Vehicles, and Transportation Network Companies

Taxis and hired vehicles are often the only source of transportation in areas not serviced by public transportation systems and during times that such systems are not providing service (e.g., late at night). The difference between taxis and hired vehicles is that the latter can be hailed on the street or at queues. But, the accessibility issues are similar. The lack of wheelchair-accessible taxis has been a serious problem for people with disabilities. But, accessible taxi vehicles exist and have been on the market for many years. The most well known in the U.S. is the MV1 (2016), manufactured by AM General. The MV1 is popular with taxi companies that target people with disabilities as customers. An example is VETaxi (2016) in Pittsburgh, which takes the added step of employing veterans and military family members as drivers. Some cities have fully accessible taxi coverage, like London, where all new black cabs are wheelchair accessible. Some also have hearing aid induction loops. The black cab is also a purpose built vehicle, designed specifically for London but it has been adopted in other cities (London Taxi Company, 2016). Car manufacturers and aftermarket customizers have produced many models of accessible taxis adapted for wheelchair use. Since accessible vehicles designed for taxi service are available and they are being used in many cities, why aren't all taxis accessible?

New York City's recent experience provides some insight into this problem. The New York City Taxi and Limousine Commission licenses and regulates over 50,000 vehicles, including New York City's medallion (yellow) taxicabs, and is the most active taxi licensing regulatory agency in the U.S. (City of New York, 2016a). The

Commission specifies the models of vehicles that can be used for taxis in the city, a practice that is not uncommon worldwide. In 2009, the Commission ran a competition to select the vehicle designs to be used for the next 10 years (City of New York, 2016b). One company offered a universally designed vehicle. Accessibility would have been a universal feature in all their vehicles, had they been the successful bidder. This vehicle was no larger than its competitors. It had an adaptable seating configuration and rear ramp that provided space for wheelchair users when needed, and, when not carrying a wheelchair user, it provided the possibility of extra luggage space. However, the Commission selected a vehicle which was not equipped for universal accessibility; although a custom adaption was provided. The decision to avoid making every taxi accessible resulted in a class action lawsuit against the Commission, which was recently settled, mandating that half the approximately 13,500 cabs be wheelchair accessible by 2020 (Brown, 2014). Currently, only 600 wheelchair-accessible taxis are in service (Figure 7.1).

As the example indicates, taxi and hired car services are governed by local laws and regulations in the U.S. as they are in many other countries. The question of how they are covered by the ADA in the U.S. is a matter being decided in the courts. Advocates argue that Title III of the ADA requires accessibility to all "public accommodations" and view taxis as such. Operators argue that the *service* is a public accommodation, not the vehicle, and that they only need to have enough accessible vehicles to meet demand available. At present, the law is being interpreted in favor of the operators, as in the NYC case. How many accessible vehicles, what kind, what features, and to what degree each taxi operator needs to provide them, is still not settled law.

TNCs are successfully competing with taxis in providing on-demand transportation. Their crowdsourcing approach means they can recruit large numbers of drivers quickly with no capital outlay for vehicles. This leads to response times far superior to taxis which traditionally have used a dispatcher to schedule trips. Further, each individual operator is rated by the rider so that the business model promotes good customer service. Taxi companies are adopting similar Internet-based hailing and rating services like Curb (Verifone Transportation Systems, 2016), but they are inherently at a disadvantage over crowdsourcing approaches since the numbers of drivers and vehicles are limited. Some local governments have made TNC services illegal and others have prohibited service to some locations (e.g., airports). The argument for such restrictions is that taxis are licensed and must adhere to rules and regulations governing vehicle safety and operator

Figure 7.1 The MV1, an accessible taxi vehicle.
Source: Photo courtesy of Mobility Ventures: Photographer: Rob Wurtz

honesty and integrity. Further, it protects taxi licensees, often small business operators, who must invest significant amounts of money to purchase a license.

However, the TNCs maintain that they are not covered by the ADA at all since their drivers operate their own vehicles and are not employees. Disability advocates argue that without coverage by regulations, discrimination is likely. Further, if competition from TNCs drives taxis out of business, it will seriously decrease the availability of accessible vehicles for hire unless the TNCs provide accessible vehicles. Competitive pressure from TNCs is cited as one of the reasons the number of accessible vehicles in San Francisco has rapidly decreased as taxi operators seek to reduce costs (Kwong, 2014). The status of TNCs under accessibility laws like the ADA is still in flux as several lawsuits and policies governing TNCs across the country proceed (Cowan, 2014). While there are efforts to require TNCs to file accessibility plans, to maintain a fleet of accessible vehicles as a percentage of all vehicles, and to train their drivers on best practices for serving people with disabilities, regulations and laws at the local, state, or federal levels are still in early stages of development (e.g., NCSL, 2014).

Uber and Lyft report that they have internal policies in place to support riders with disabilities which, in some cases, are influenced by negotiated agreements with states and local municipalities (e.g., Portland, California, etc.) (see Figure 7.2). Even though TNCs may take the position that they are not covered by the ADA, in negotiating with local municipalities, they agree to offer services to riders with disabilities to gain access to a market and promote goodwill in the community. Their policies toward people with disabilities are also part of their emphasis on inclusion from a social branding perspective. Both companies explicitly prohibit discrimination against people with disabilities. Violations of these policies may lead to account deactivation. Lyft's policy includes explicit details on service animals and stowable wheelchairs, along with advice for drivers on how to support such customers and emergency phone numbers for case assistance (Lyft, 2017). Their policy also covers drivers with service animals.

Figure 7.2 Accessible Uber vehicle.
Source: Photo courtesy of Uber

Shuttle Buses

Shuttle buses have been used for years to ferry people from transportation terminals to hotels and rental car agencies, and to transport passengers within large transportation complexes like major airports. They are also being used for "end of the line" service on fixed route transportation corridors to extend the service areas. These services utilize many different size vehicles, including seven- to nine-passenger vans, 12-passenger small buses, and full-size buses, depending on demand. In some rural low-income areas, shared small buses and vans are the only transportation services, operated by local entrepreneurs or non-profit organizations like independent living centers or senior centers. Many shared bus and van services are operated on a regular schedule and others are operated in a hybrid of on-demand mode and scheduled service with additional trips added as needed. While some serve only specific destinations, others travel a regular route and pick up people along the way at informal stops or by hailing. This type of service is often called "jitney" service.

As public accommodations, shuttle services should certainly comply with accessibility laws but, unlike public transit agencies and taxis which are already closely regulated by government agencies, and TNCs, which are increasingly subject to regulation, there is no existing way to regulate and enforce compliance with the laws other than individual complaints and lawsuits under accessibility laws governing public accommodations. An example of how this works is a case against ANC Rental Corporation, and its subsidiaries, Alamo Rent-A-Car LLC and National Car Rental System, Inc. In this case, the defendants reached a settlement agreement with the Department of Justice to resolve complaints filed by travelers who use a wheelchair or scooter regarding the lack of accessible airport shuttle buses (U.S. Dept of Justice, Civil Rights Division, & United States of America, 2004). As part of the settlement, Alamo and National agreed to provide at least one accessible shuttle bus at airport locations, provide equivalent, accessible pickup and drop-off services when shuttle service is not available, and purchase accessible vehicles with a capacity of 17 or more passengers. While a settlement does not become law, when the terms are public, it sets a precedent in a particular industry.

Services based on small buses have a good future in accessible transportation. In particular, as the population ages, it will be necessary to increase service to low-density neighborhoods like those in suburbs and, in older cities, deep into neighborhoods with narrow streets. Many countries are already providing such service to dense urban neighborhoods. Even where they fit on the streets, residents resist the use of large buses in quiet neighborhoods and streets because they are visually and audibly intrusive. Further, where demand is low, small buses are much less expensive to purchase and operate.

Up until recently, accessibility to small buses and vans was provided by a lift. Today, both ramps and lifts can be found since small buses are now available with low floors and folding ramps. Reports from service providers, and limited research by the first author and his colleagues, demonstrate that the low-floor models with kneeling features and ramps provide more inclusive access than the lift buses. Many people who are ambulant and those with sight and grip limitations find it difficult to use the stairs on high-floor lift-equipped buses. Further, wheelchair users do not have to negotiate the tight turns between wheel wells at the front of large low-floor buses because the entries are behind the wheel wells in the smaller buses. However, there are still some problems that need to be worked out in the smaller buses. They are more costly and some individuals may need assistance negotiating ramps. Further, they are unfamiliar to riders with visual impairments, who cannot tell the difference between them and the more familiar large buses or high-floor buses, on which they would use the stairs (Figure 7.3a and 7.3b).

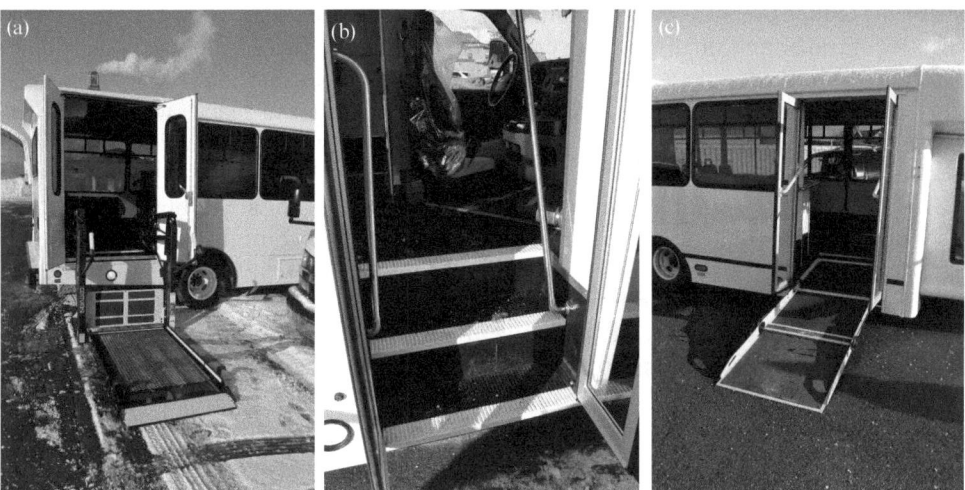

Figure 7.3 Shuttle buses. (a) High-floor shuttle bus with lift, (b) high-floor shuttle bus with stairs, and (c) low-floor shuttle bus with ramp.
Source: IDeA Center

Small buses can be operated using TNC models. For example, Bridj (2015) used market demand within neighborhoods to dynamically identify and generate virtual bus stops for service between the areas. It was designed to provide efficient and affordable transportation for people who frequently travel from one specific area (e.g., neighborhood) to another (e.g., office district). Riders indicated their location and travel desires through a smartphone app, which also provided details on stops and routes. Small transit vehicles were dispatched using optimization techniques. Service animals were always permitted and Bridj asked people with disabilities to contact customer service in advance. Bridj used vehicles provided through local subcontractors, which means that the degree of accessibility was based on the number and type of accessible vehicles in the subcontractors' fleets. Bridj had an interest in providing more accessible vehicles but they had many challenges in implementation of their services during the early stage in the company's existence, and ultimately failed (Figure 7.4).

The Bridj model naturally finds and fills gaps within existing transit systems and supports spontaneous travel. Further, the virtual bus stop concept can reduce the length of

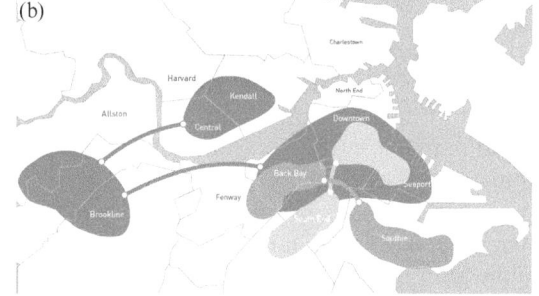

Figure 7.4 (a) Bridj vehicle, (b) service area map.
Source: Photo courtesy of Bridj (www.facebook.com/TakeBridj/photos; www.bridj.com/blog/2015/3/31/expanding-smarter-mass-transit-in-boston)

the pedestrian trip to the bus stop for a rider with a disability. As such, this model may be able to reduce pressure on STS demand. Virtual bus stops also allow for adjustments when local traffic and infrastructure create barriers. Since there is no fixed roadside infrastructure, it is possible to shift the stop to a more usable location. However, accessibility of stops may be compromised if the stops are located in places without parking spaces large enough to deploy ramps or lifts to curbs or, where street loading is required, lack curb ramps.

Skedaddle (2015) is another variation of the TNC model for small bus service. It focuses on the longer, charter trip market. Once 15 riders have committed to a trip, the bus is scheduled. The service requires reservations multiple days in advance but may be useful for supporting less regular, longer routes for communities that can make such plans. For example, Skedaddle may be attractive for veterans, military families, and seniors since origins and destinations from suburban and rural bases or senior communities become more cost-effective than taxis or TNCs.

Sharing Cars and Rides

The most common form of shared cars is the traditional rental car. Rental car agencies are public accommodations and must comply with laws like the ADA. They are required to make accommodations for drivers who may need adaptive controls, by providing or renting the controls, or allowing drivers to install their own. Additional accommodations established through settlement agreements between rental car companies and the Department of Justice include providing accessible shuttle services, allowing individuals who are visually impaired to be able to rent a car when accompanied by a licensed driver, and allowing persons with disabilities the "use of their service animal" (Florida Alliance for Assistive Services and Technology, 2012).

New technologies are changing the rental car market. Shared cars using a subscription payment model enabled by mobile device apps like Zipcar allow riders to use the cars for much shorter periods of time and leave them at designated locations around a community. This eliminates the location-based model of rental cars that concentrated cars at depots, usually at airports or other major transportation terminals. This new model is likely to transform the rental car industry, a trend not lost by a major rental car operator, Avis, which purchased Zipcar and operates it as a subsidiary. Lawsuits were filed against Flexcar and Zipcar in 2007, plaintiffs alleging multiple ADA violations including a lack of hand controls available for vehicle use, company restrictions prohibiting the transport of animals (including those designated for assistance), and rules prohibiting car share members to be driven by a personal care aide (Equal Rights Center, 2007).

Drivers can now use smartphone apps to share rides. Examples include Commuter Connections, Carma, and RideScout. Commuter Connections (National Capital Region Transportation Planning Board, 2015) makes it easier to directly identify and coordinate free carpools. Commuters can search for carpool partners by entering their home and work locations and work hours and receive contact information for commuters with similar schedules. Information on the location of park-and-ride lots where commuters can meet carpools or vanpools is also provided. The app even allows access to the guaranteed ride-home program as well as carpool rewards. Carma (2015) is similar but is designed for drivers to receive distance-based payment from the passenger. Drivers can also designate passengers who can ride with them for free. While most of the Carma cities are major urban areas, the service is also deployed in smaller cities (e.g., Cork, Ireland). Both of these systems are useful for populations who are having difficulty finding good carpool options. Commuter Connections' guaranteed ride home feature is also important for those who lack the financial resources to switch to a taxi or TNC when a carpool plan unexpectedly fails. TNCs are also moving into the shared ride business. Lyft Line and Uber Pool, for example, are applications that allow passengers to share rides if they are heading in the same direction. This can reduce the cost of Lyft or Uber rides significantly.

Autonomous Vehicles

Fully robotic vehicles, or "autonomous vehicles," are still in the development phase but they have the potential to dramatically alter the future of on-demand transportation. The highly publicized Google Car team (now called Waymo) has illustrated the potential power of autonomous vehicles to provide an accessible form of transportation. Steve Mahan, a blind man, demonstrated the power of autonomous vehicles in a demonstration video in which he was shown getting a ride to a nearby restaurant for drive-through tacos and a trip to the dry cleaner (Google, 2012). In this clip Mahan remarks that having a self-driving car can give him the autonomy of doing things for himself on his schedule. Subsequent demonstrations show other people with disabilities experiencing rides in Google's custom built autonomous vehicles (Google, 2014; Liljas, 2014). These new vehicles lack driver controls, so manual controls are not in the way for people entering the vehicle, and the ability to drive is not required. This should lead to more accessible vehicle designs for individuals who can transfer in and out of a vehicle on their own as long as the vehicles are not too small. Proposals for very small robotic cars will make it impossible for many people with mobility impairments and older people with a range of motion limitations to enter and exit (Figure 7.5).

Figure 7.5 Google autonomous vehicle. This vehicle is too small for many people to exit and enter.
Source: Photo courtesy of Google

Start-up companies are being launched to apply autonomous vehicle technologies to shuttle bus service. EasyMile (http://easymile.com/) is already manufacturing a small driverless shuttle bus vehicle called the EZ10, with trial installations underway in nine cities around the world. Local Motors has introduced the Olli concept shuttle vehicle (https://localmotors.com). Both vehicles are designed to be used in regular scheduled shuttle service and in on-demand shared ride service. Local Motors has adopted two advanced technologies. The first is a "co-creation" software platform where anyone can contribute ideas and participate in the design process. This provides an opportunity for people with disabilities and experts in rehabilitation to engage directly in the design process and even own their ideas. The second is advanced manufacturing using digital fabrication at local "micro-factories." The company has produced the first "3-D Printed Car" in a demonstration of their manufacturing technology. A general goal of the company is to institute a business model that reduces the development cycle of new vehicles to very short time frames. Both vehicles are low-floor designs and the EZ10 includes a "kneeling feature" and telescoping access ramp (Figure 7.6).

Although there are many challenges to the safe operation of fully autonomous vehicles, these challenges can be reduced by applying some constraints. For example, trips along prescribed routes are much easier for autonomous vehicles to manage than driving with varied route selection. The CityMobil2 project (2015) is exploring this type of application. There is already anecdotal evidence that people with disabilities are utilizing CityMobil2 demonstrations for improved quality of life. Many of the legal concerns in deployment of autonomous vehicles are reduced when autonomy occurs within "less" public areas. The TARDEC ARIBO program is exploring neighborhood autonomy within U.S. military bases and other private campuses (Brooks, 2015; O'Connor, 2015). The Ft. Bragg

Figure 7.6 Exterior of EZ10 self-driving transport vehicle with ramp deployed.
Source: Photo courtesy of EasyMile

deployment is especially interesting as the system will traverse routes valuable to a wide range of users, thereby permitting a better understanding of how autonomous vehicles and novel service models operating in a mixed mode setting will be accepted by the general public. The route will include a connection to external transit service for multi-modal trips. Aside from technical and user issues, this research project will also include performance and cost evaluation. This should help when estimating impact for similar systems funded through non-military sources.

There are many opportunities to improve service delivery to groups currently under-served or challenged by public transportation using autonomous vehicular systems in an on-demand service model. First and foremost is the possibility for greater service area coverage and density. With the elimination of costs and logistics associated with drivers, it becomes possible to scale-up on-demand transportation services. Current taxi driv-ers optimize their driving and scheduling strategies for personal gain while a driverless system could be optimized for societal gain. This could lead to better coverage and ser-vice in areas that typically lack high-value rides or at times when demand is so low that running vehicles on a route is cost prohibitive. Another important element is the ability to use vehicle designs that are more open and easier to board. Removing the driver seat supports new, innovative vehicle designs. Driverless electric vehicles are likely to have a significant positive impact on trip cost (Burns, Jordan, & Scarborough, 2013). Estimates on the savings from even a modest impact in the U.S. are upwards of $50 billion per year (Burns, 2013) and 2/3 of the current cost of a taxi ride (Fagnant & Kockelman, 2015). One study that explored the impact of autonomous driving on seniors and people with disabilities who are currently unable to drive found overall vehicle miles traveled could increase by 12 percent due to newfound travel independence (Harper, Hendrickson, Mangones, & Samaras, 2016).

Partnership Models

It is not uncommon for STS or special transport services to have partnership agreements with taxi companies. Such agreements are common in many cities and other cities are developing partnership agreements. A taxi ride is much less expensive than an STS ride so individuals who do not need wheelchair lifts or ramp access, can often be accommo-dated more economically using taxis. In some agreements, taxi drivers receive vouchers or credits from the public transport agency for providing such rides. In others, the tran-sit agency may simply pay for the ride to reduce their costs. For example, Washington, DC, pays the entire cost of a taxi ride for an STS-eligible rider (Lazlo, 2016). Boston's transit agency provides a service that allows eligible riders to hail an accessible taxi and pay with a debit card the agency provides. For every $2 spent with the card, the agency reimburses the taxi company $13 but this cost is still far lower than an STS ride provided by the agency (Young, 2015). Partnership arrangements like these can provide less ex-pensive service and eliminate the need for advance reservations, but it is important that taxi drivers are trained to address the needs of people with disabilities and that sufficient accessible vehicles are available to meet demand.

Both Uber and Lyft partner with local transportation providers who have wheelchair-accessible vehicles. Some of these providers are non-government, non-profit organizations. The TNCs refer riders to these providers. In some cases, riders are required to make requests 24 hours in advance which is similar to regular STS. Uber is interested in how their service model compares to traditional STS but notes the common challenges of limited wheelchair-accessible vehicle supply. Transit agencies are developing agreements with TNCs as well as taxi companies (Young, 2015). The opportunity to increase income from transit agencies may be an incentive for TNC drivers to obtain accessible vehicles.

TransitCare, in the Brisbane, Australia, metropolitan area has developed a social service approach to transportation to ensure that individuals with disabilities received the most appropriate form of services. Fares can be paid by individuals, reimbursed by third-party payers, or subsidized through the public transit agency. They have their own accessible small buses and have established cooperative arrangements and integrated scheduling with local public transportation agencies and taxi operators (TransitCare Ltd, 2015). They also recruit and coordinate volunteer drivers who use their own cars to transport riders.

TNCs have an opportunity to partner with STS service to improve their effectiveness by overcoming the problems identified earlier in this chapter. Not only might they take on the booking arrangements but also use crowdsourcing to triage service to the best available service provider for the individual. One of the most important benefits could be reducing the lead time for ordering a ride by shifting STS rides for non-wheelchair users to regular vehicles driven by taxi or TNC drivers. Using the data collected from their ridership and drivers, they may also be able to identify ways for STS providers to operate more efficiently and provide better service. For example, Uber has received positive feedback on the cashless nature of the payment model from riders who are blind. This helps mitigate conflict and increases safety among the participants. Likewise, Lyft has worked to ensure VoiceOver capability in iOS for a better screen reading experience. Both companies cite the value to real-time arrival alerts for riders.

Working with an organization serving the deaf community, Uber has recently updated their driver user interface to better support drivers who are deaf or hard-of-hearing. User interaction elements include: alerting the rider that their driver is deaf or hard-of-hearing, encouraging the rider to identify their destination in the app rather than speaking it to the driver, better non-audible notifications, and removing the option to call the driver when matched with a deaf or hard-of-hearing driver. Uber has also incorporated visual and vibration features into the app to better support riders who are deaf and they have received positive feedback regarding the ability to enter a destination within the app. Lyft also actively recruits drivers who are deaf and hard-of-hearing, with good results (Jaffe, 2014; Said, 2015). Uber has also reached out to veterans as a source of drivers through UberMILITARY (Gates & Kalanick, 2014), which has a focus of adding 50,000 veterans, service members, and military spouses as Uber drivers. Recent statistics indicate that over $35 million has been earned by drivers in the program (Uber Newsroom, 2015). The program also includes local chapters and a Family Coalition to support drivers in the program.

At the time of this writing, Uber was pilot testing partnership models designed for accessibility in multiple urban areas, including larger cities like New York and Chicago, with the plan to develop data-driven best practices. In Chicago, San Diego, and Houston, Uber was testing a pilot called UberASSIST where top drivers receive additional training developed in partnership with groups like the Open Doors Organization. This training includes core topics within disability terminology, legislation, methods of assisting specific disability groups, and technical aspects of assistance like training on transfers and wheelchair breakdowns. This sort of specialized training within the larger Uber driver pool is in line with requests from disability rights groups. Lyft has similar pilots underway.

Payment Technologies

Fare payment is an important aspect of on-demand service. It is particularly important in partnership models where there may be several different service providers working together and subsidies are involved. On-demand services often have complicated fare calculation models that are difficult to understand due to variables such as zones, travel time, surcharges for rush hour, baggage and extra riders. As noted early in this chapter, STS can require special payment methods for accounting reasons that are aggravating and burdensome, especially for low-income people. Another major problem is lack of continuity across systems in a local area. Public systems can also have very different eligibility requirements based on age, income, location of residence, and disability. Many require pre-authorization to be eligible for service. But current information technologies offer solutions that could simplify fare payment significantly. In fact, the success of TNCs is partly due to their simplified payment model which automatically calculates fares, provides estimates of fares in advance and allows the user to manage their payments and expense accounting easily.

Many cities have instituted comprehensive smart card systems that work the same on all modes of service with multiple providers. Some of these systems also function as a cash card for other services. The T-Money system in Seoul, for example, can be used at a variety of purchase points, like taxi cabs, convenience stores, etc. Scaled-down versions of this functionality are also present at some university campuses in the U.S. Being able to travel and purchase items with only a smartphone, NFC-equipped smartwatch (e.g., Apple Watch), or smartcard can have significant value to people with disabilities due to reduced cognitive load when managing multiple accounts and easy identification of what card (if any) is used to pay. The reduced physical and sensory challenges of these systems increase convenience and personal autonomy. In some cases, the on-board security of a device may also help protect against theft and unauthorized use.

Although the number of people who use smartphones is increasing rapidly, there is still some resistance to adopting payment systems that rely on establishing a balance on a smart card. Some advocates argue that many people with disabilities do not have the resources to maintain a balance or buy an expensive phone. However, the adoption of these new technologies benefits many others with disabilities as well as the general ridership. As noted above, there are solutions that can overcome these problems. For example, Washington, DC, simply issues debit cards that STS-eligible riders can reload on their own to pay for taxi service. Crowdsourcing approaches, however, rely on smartphones, which are still not affordable for many people with disabilities.

Problems

- There is no common definition of DRT. Definitions developed for regulatory purposes obscure the full range of services available.
- "Silos of service" separating fixed route from DRT and public and private services cause inefficiencies that translate into access limitations.
- The high cost of STS is a growing burden for public transit agencies and service is often of poor quality.
- The legal responsibilities for accessibility in taxi service, TNCs, and other demand responsive services are not well developed and may vary significantly from place to place.

Recommendations

- Utilize the many available solutions for making DRT accessible.
- Engage in systematic research rather than rely on anecdotal or descriptive information on successes or limitations of services.

- Explore how new developments in mobile computing and information technologies like crowdsourcing are providing challenges to existing DRT services like taxis but also enabling new forms of service and partnership models.
- Consider whether or not new forms of payment, like smartphone apps, or smart cards are barriers or facilitators of accessibility and for whom.

Directions for Future Work

- Improving the usability of fixed route service beyond minimum mandates can reduce the burden to public transit agencies by reducing dependence on STS.
- Methods to integrate accessible on-demand transportation services like STS, car sharing, and taxi services with fixed route service are needed to ensure a comprehensive and cost-effective approach to community transportation.
- Explore the advantages of small buses and new information technologies for providing accessible and convenient service to all riders—STS for all.
- Explore how autonomous vehicle technologies can improve accessibility and usability of public transit by providing more demand responsive services for people with disabilities.
- Revise regulations to address partnership models and integration of accessible DRT transportation with systems serving the general population.
- Develop partnerships between developers of innovative services and disability advocates to ensure that new services will not discriminate.

Source: IDeA Center

8 Paratransit Scheduling and Routing

Zachary B. Rubinstein and Stephen F. Smith

Overview

As noted in Chapter 7, paratransit service is a major economic burden for public transit agencies, especially when budgets are not increasing and costs are rising. Although universal design throughout the Travel Chain can reduce passenger reliance on paratransit, the aging of the population will increase demand. So, agencies must seek ways to reduce this burden by improving operations without reducing service quality. More effective scheduling and routing systems can go a long way toward improving service quality and reducing the problems noted in the last chapter.

Paratransit service is notoriously difficult to provide efficiently, cheaply, and predictably. The problem of generating daily schedules that meet customer service constraints while minimizing the number of vehicle hours on the road is inherently complex, and represents one major challenge. An even greater challenge, however, is effectively managing these schedules (and operations) in the face of unpredictable execution circumstances.

Unexpected traffic congestion, vehicle breakdowns, new client pickup requests, driver call-offs, and a myriad of other events continually cause disruptions to the schedule as it is executed, requiring the ability to accurately monitor progress, project and analyze the downstream impact of current events, and take rescheduling actions as necessary to keep operations moving in the best possible direction.

Historically, human dispatchers have had to try to meet these challenges through largely manual ways, using paper and radios, and under these circumstances even expert dispatchers can become overwhelmed and forced to take myopic decisions in order to keep pace with events. However, as computers have become more prevalent in the industry, software and hardware technologies have emerged that provide better tools for meeting these challenges. Mobile, in-vehicle devices now enable real-time electronic tracking of paratransit vehicle status. The recent development of technologies for dynamic scheduling now provide mechanisms for using vehicle status information to provide early warning of delays, and for generating rescheduling options that minimize the impact of disruptive events on overall performance.

This chapter examines the challenges associated with the paratransit scheduling problem, and surveys the current state of the science in terms of technologies and tools for solving them.

The Scheduling and Routing Problem

The scheduling problem faced by paratransit services, at its core, involves the assignment of client requests to vehicles. In the most basic case, client requests are negotiated in advance, and a client request specifies a pickup time, a pickup location, and a destination (or drop-off location). Depending on the policies of the service provider (and in some cases government regulations), the negotiated pickup time is typically expanded into a *pickup window constraint* that must be satisfied. For example, a pickup window that starts 10 minutes before and ends 20 minutes after the negotiated time might be specified. In addition to this pickup window constraint, there is also a *ride-time constraint* that limits the maximum time that a rider can be on a vehicle. This constraint is usually specified relative to the expected travel time from the pickup location to the destination. For example, the ride-time constraint might specify that the client cannot ride for longer than the lesser of 2 hours or twice the expected duration to go directly from the pickup location to the drop-off location, implying that if a trip from location A to B takes 25 minutes, then the client should not be on the vehicle any longer than 50 minutes. For service providers that accept client requests in real-time (discussed further below), a third *pickup delay constraint* (e.g., pickup within 45 minutes) is substituted for the pickup window constraint. In addition to these pickup and travel time constraints, which reflect client quality of service, assignments are also subject to the numbers and types of paratransit vehicles that are available, and their usage constraints. Different vehicles have different capabilities (e.g., ambulatory, wheelchair transport) and carrying capacities (sedan vs. van) that must be matched to client needs. An efficient paratransit schedule will result in fewer vehicle hours on the road while still meeting the service times that were either negotiated with the client or mandated by regulations.

Variants of the above pickup and delivery problem have been studied extensively in the academic literature (Berbeglia, Pesant, & Rousseau, 2011; Cordeau, 2006; Cordeau & Laporte, 2007; Jain & Van Hentenrych, 2011; Lim & Zhang, 2007; Nagata & Brysy, 2009; Rubinstein & Smith, 2011; Rubinstein, Smith, & Barbulescu, 2012), and a broad range of solution methods have been proposed. For the most part, this work has focused on the "static" optimization problem where all requests are known in advance; however, in recent years more attention has been given to dynamic, continuous scheduling problem variants, where client requests arrive incrementally over time. To examine the relevance of this work to currently available and emerging paratransit scheduling technologies, we first consider the principal scheduling challenges currently faced by paratransit service providers.

Paratransit Scheduling Challenges

Most paratransit organizations require that requests be made 24 hours in advance of the trip. These requests are a mixture of *subscription trips* (i.e., trips that repeat periodically, such as a weekly therapy session) and regular *one-time trips*. Some number of dynamic, *will-call trips* is also supported; these are same-day one-time trips and are typically used for transporting a client back from a destination after an event of uncertain duration (e.g., a doctor's appointment). Some organizations are starting to consider broader support for same-day reservations (a point we will come back to later).

Given these sets of service demands, a typical paratransit service provider's planning cycle starts anew each evening with the generation of the next day's schedule. The requests that have been received up to that point as well as current information about driver availability and other vehicle usage constraints is gathered, and the first major challenge is to generate an efficient schedule that meets the constraints described above. Generation of a good fleet schedule is an inherently combinatorial problem; there are many possible schedules for servicing a given set of requests, some obviously better than others with regard to efficiency, cost, and quality of service. The possibilities can be overwhelming if the schedule is built manually, and lead to poor schedules and service quality. To further complicate matters, the set of demands themselves are uncertain and in certain circumstances (e.g., bad weather days), the cancellation rate can be as high as 15 percent. Service providers are caught in a bind as to whether or not to provide sufficient resources (vehicles and drivers) to meet all requests, which is costly, or to over book vehicles in the schedule in anticipation of cancellations.

Managing the daily schedule through its execution over the coming day is a second major challenge. As the day starts, the initially scheduled vehicles depart and begin making their stops. At this point, dispatchers begin to monitor the execution of the schedule and adjust it when the inevitable divergence happens due to some unexpected event. Monitoring the execution of the schedule depends on the infrastructure available to the drivers and dispatcher. At a minimum, drivers report back to dispatchers when drop-offs and pickups are made. Each update represents a potential decision-point for the dispatcher. If the stops are on-time, then no change is warranted. But, if they are delayed, then the dispatcher has to decide if alternative means are needed to handle any downstream requests that are in jeopardy of being late. Note that determining potential downstream effects requires the propagation of delay forward through the vehicle's route, which, over all routes, can be taxing if left solely to a human dispatcher.

If a trip must be rescheduled, then alternative vehicle options must be identified for servicing the problematic trip(s). This complex activity requires dispatchers to understand all route manifests and their current state of execution. Then, the dispatcher must determine whether or not a pickup and drop-off can be added somewhere to the route without violating any constraints associated with the requests currently being served by that route. Even more complex, some of the best alternatives for rescheduling a problematic trip may require moving several downstream trips among routes. Some unexpected events, such as a vehicle breakdown, require rescheduling of multiple requests and further stretch the dispatcher by forcing the simultaneous computation of multiple options.

Will-call requests, which are not built into the initial schedule, represent another class of unexpected event that must be handled through the day. Although one might consider strategically inserting slack into the initial schedule to accommodate these projected events, the inherent uncertainty associated with the timing of these events makes this a difficult proposition and this approach is rarely adopted. Instead, the dispatcher will typically make a real-time decision as to which vehicle to divert based on the current location and expected route of each of the vehicles. These decisions could similarly benefit from an ability to explore options for rearranging existing assignments. In practice, dispatchers typically either allocate the vehicle that is physically the closest or solicit interest from current drivers.

Note that this type of reasoning is also what would be necessary if same-day reservations were allowed. If this were to become the case, the increase in the number of such requests would undoubtedly add significant additional time pressure to human dispatchers' decisions.

In addition to managing the schedule to address problematic and will-call situations, dispatchers can also look for opportunities to reduce costs. For example, if the number of cancellations is high, then it might be possible to consolidate multiple routes and send one or more drivers home early. This type of opportunity is difficult for human dispatchers to recognize. They have to notice that there is slack on multiple routes and that there is an opportunity to free up a route (or routes) by moving its remaining trips to the other routes that have slack.

Overall, the decision making necessary to maintain efficient operations in such a dynamic environment is extraordinarily complex and requires highly skilled people, capable of tracking many details and projecting future consequences over time. Even then, there is a limit to what they can achieve without computational support. In the next section, we describe the types of technology available and the range of ways that they can support this planning and re-planning process.

Emerging Technology

With the advent of automated scheduling tools, GPS tracking systems, and mobile connected devices, there has been a significant move to increase efficiency in managing paratransit fleets through better scheduling and improved real-time situation awareness. A number of technical solutions exist or are becoming available, ranging from interactive systems that provide different levels of assistance to human dispatchers to nearly or fully autonomous systems. In this chapter, we examine the new opportunities for performance improvement that these systems offer.

Increasingly, paratransit service providers are incorporating the use of automated scheduling tools into the process of generating the next day schedule. There are a variety of commercial products available that provide this capability. In most cases, these systems provide solutions to the static version of the problem (i.e., given a set of input requests, generate a set of vehicle itineraries that achieve them), drawing on techniques and heuristics that have been derived from the academic research in this area mentioned earlier. For the service provider, the ability to automatically generate a next-day schedule addresses the complexity and scalability issues associated with manual schedule development. Auxiliary capabilities that allow human planners to interactively adjust the schedule provide the necessary flexibility to address specific idiosyncrasies (e.g., pairing a specific driver with a specific client). The result is a baseline schedule that provides a more efficient starting point and reference for managing the next day's operations. The same automated scheduling capability with a slight extension to incorporate the current execution state of each vehicle can also be reapplied periodically through the day to integrate new requests or otherwise "reset" the schedule to match changed circumstances. A much smaller number of systems also provide the separate capability to incrementally integrate new requests into the schedule.

Technology support for managing and executing a schedule, at the most basic level, implies an ability to monitor progress, and to this end paratransit service providers have increasingly introduced tracking systems into their operations in recent years. Typically, vehicles are equipped with GPS-capable devices, such as smart tablets, that communicate (usually via wireless cellular technology) with a backend tracking system in the operations center. The GPS capabilities allow for AVL tracking of the vehicles so that the system is aware of where vehicles are at any time. The in-vehicle device interface is designed to present the current itinerary to the driver and to capture information from the driver such as client pickup and drop-off times, progress updates, and incident reporting. The backend tracking system records this status information as it is received and makes it available to the dispatcher (e.g., through map-based visualizations, superimposition over the reference schedule, etc.).

The emergence of such "connected vehicle" tracking systems also provides the infrastructure for much more sophisticated forms of automated scheduling support. At one extreme, there are technology solutions (e.g., Ecolane, 2015) that advocate full system autonomy (i.e., taking the dispatcher out of loop). Under such an approach, the schedule is typically generated and executed in a *rolling horizon* fashion, i.e., a new schedule is generated periodically (e.g., every 30 minutes) using current vehicle location and state information and then executed in open loop until the next period's schedule is installed. The strong assumption that is made in this type of approach is that the automated system's model of the target domain includes all constraints that could possibly impact vehicle schedules (or can be made to incorporate all relevant constraints over time).

Recent research in dynamic scheduling (Barbulescu, Rubinstein, Smith, & Zimmerman, 2010; Rubinstein, 2002; Rubinstein & Smith, 2011; Rubinstein et al., 2012; Smith, Becker, & Kramer, 2004) advocates a more adjustable approach to autonomy. This work has focused on the development of continuous scheduling frameworks, where schedules are incrementally extended and revised over time in response to execution dynamics. These frameworks center on the notion of a *live schedule*, a virtual model of future events as prescribed by the current schedule that is continuously updated as new execution results become known. The live schedule also includes a representation of the constraints underlying the schedule (e.g., pickup windows, client load/unload times, pickup/drop-off sequences, ride-time constraints, etc. in the paratransit domain), which allows early alerting of future problems (e.g., a pickup that is projected to be late due to vehicle travel delay). Incremental search procedures are used to provide a companion capability for revising the schedule in response to detected problems in a controlled manner that re-optimizes future scheduling decisions while attempting to minimize overall change. This dynamic scheduling approach has proven effective in a number of complex pickup and delivery problem domains and has recently been applied to the paratransit scheduling problem (Rubinstein & Smith, 2011; Rubinstein et al., 2012).

Although it is possible for a dynamic scheduling approach to operate autonomously, it also offers a rich basis for interactive decision-support in application contexts where it is not practical to assume that the scheduler's model can sufficiently capture all relevant aspects and details of the real-world problem. First, it provides a general focus of attention mechanism. As execution results are reported, the propagation of consequences through the live schedule can inform dispatchers of future events that require their attention. Notification that a drop-off task has just been completed, for example, can lead to identification of a future pickup that is likely to be late. Likewise, cancellation of a request or a "no show" pickup attempt will free up vehicle capacity that may suggest some rebalancing of allocated requests. By limiting how far into the future that the system looks for potential problems (and opportunities), distant events that may resolve themselves as execution plays out and do not require immediate attention can be filtered out and ignored.

A second broad decision-support capability provided by dynamic scheduling technologies is option generation. Given an identified problem (e.g., a future pickup that is expected to be late according to the current schedule), incremental search procedures can be applied to generate a number of possible corrective actions (e.g., request reassignments) that the dispatcher might take. One basic advantage of an incremental change framework (as opposed to an option generation scheme that regenerates a new schedule from scratch) is that the options generated tend to be less disruptive and hence they are easier to comprehend and evaluate by human dispatchers. Simple options may examine simple moves or exchanges of requests across vehicles. However, better solutions can often be found if more extensive, multi-move changes are considered. Such options, which are typically not considered by dispatchers due to their complexity, can also be generated quite efficiently (e.g., Lim & Zhang, 2007; Nagata & Brysy, 2009; Rubinstein et al., 2012). Similar sets of options can be generated to integrate new will-call requests into the schedule as they are received.

Broader Applicability

Although we have focused above on the problem of managing daily paratransit operations, the automated scheduling technologies we have discussed have broader applicability to other business functions of a paratransit service provider. In this section we briefly consider two: negotiating client pickup times and multi-modal trip planning.

In current practice, requests for a particular day's schedule are accumulated over days and weeks prior to the day of service. A significant challenge to service providers is that of generating options for a prospective client that both meets the specified time constraints and can be feasibly serviced together with the set of requests that have already been allocated for that day. Automated scheduling technology can make this process much easier through its ability to incrementally extend the schedule as new requests are received. For example, the dynamic scheduling capability mentioned above could be used to generate only feasible options for a client's request. Furthermore, once an option is confirmed, this same mechanism would install it into the schedule at the appropriate time, thus ensuring that the option remains viable as future requests are negotiated.

A second, related application of automated scheduling support is multi-modal trip planning. Many client requests are most economically handled by integrating paratransit vehicle service with fixed-schedule transit services. In current practice, this coordination is done manually and is susceptible to long intermediate wait times. By associating additional transit schedule synchronization constraints with the client request, it is possible to generate trip schedules that establish much tighter wait windows at trip connection points.

Problems

- To accommodate high uncertainty in predicting service and travel durations, paratransit service often builds significant slack into operating schedules at a cost to both service quality and efficiency (e.g., utilizing large time windows for pickup and ride duration).

- Most paratransit systems require ride requests a day or more in advance.
- Details of will-call and spontaneous requests are hard to predict.
- Manual scheduling by human dispatchers without computational support is difficult and has an impact on service quality.

Recommendations

- Utilize automated scheduling tools to achieve better service quality and delivery.
- Utilize vehicle tracking to improve overall performance and increase dispatcher situation awareness.
- Develop more robust scheduling tools to assist in advance scheduling and real-time adjustment of the daily schedule and maintaining service quality.

Directions for Future Work

- Extended development is needed for "live," schedule-driven demand management techniques that construct and maintain schedules from start of day through execution and provide constant, up-to-date situational awareness and rapid option generation.
- New methods are required for generating options that relax current scheduling constraints (e.g., propose a latter pickup time) in cases when it is not possible to satisfy all client constraints.
- New approaches for coordinated last mile service are needed.

Photo of a tactile map.
Source: Photo courtesy of Touch Graphics, Inc

9   Location-Based Information

Aaron Steinfeld, Anthony Tomasic, Yun Huang, and Edward Steinfeld

Overview

Digital information can be used by people with disabilities in many ways to increase independence and safety when using transportation systems. Theoretically, digital information can be configured to provide the specific content individuals need and prefer, at the right time and location. But, this goal is currently not fully realizable within public transportation systems due to technological and policy limitations. This chapter focuses on several developments in technology that have promise to advance further toward this goal.

To put the discussion that follows into perspective, it is important to understand the activity of wayfinding and how it relates to use of public transportation systems. Wayfinding involves three related tasks: orientation, navigation, and destination verification. Orientation is concerned with knowing where one is at a point in time and having an understanding of where everything else is with respect to that location. Navigation refers to getting from place to place, usually accomplished by following a set of directions. Verifying one's destination refers to knowing that one has arrived at the intended destination.

The traditional infrastructure provided for wayfinding consists of these elements:

* information on routes, schedules, and fares in the form of printed materials and large display boards;
* directional signs that guide people through stations and terminals to bays, gates, and platforms;
* identification signs that identify stops, stations, gates, platforms, and bays;
* vehicle identification information that identifies the route and direction that to which a vehicle is assigned;
* human information sources such as help lines and information desks.

Today, all of these forms of information are being digitized to a certain extent. Digital information presents new opportunities for people with disabilities as well as some potential challenges. Traditional systems do not provide continuous orientation information, which is desired by people who are blind when traversing large spaces like train station lobbies and complex intersections (Morton & Yousuf, 2011). Moreover, sign systems are only usable by people who can see and, except for some information that can be presented as ideograms, are literate in the language(s) used in the signs. Most traditional forms of information are static and thus limit content. Human-based sources of information suffer from reliability problems and frequently are only available during limited times due to cost considerations. Some traditional sources provide too much information or are so complex that they present significant challenges to interpret. Other sources do not provide enough information. Many people with disabilities, especially people with cognitive impairments and those with visual, hearing, and communications impairments, could benefit greatly from effective orientation and directional information customized for their needs. Since economic factors are driving the rapid adoption of digital technologies, it is important, during this transformative period, that people with disabilities are not excluded from using the wayfinding information resources that are provided for the general ridership of transportation systems. Rather than benefitting, they could be left to rely on expensive assistive technology or be dependent on other riders and operators. Luckily, the operators have a stake in the independence of all riders. The less they have to provide human support and the more their riders with disabilities can become independent, the lower the cost of operations.

Electronic Infrastructure

Localization, which means finding one's position in the world, within a station or vehicle is an attractive capability of wayfinding technology for many reasons (e.g., knowing which subway door a rider is exiting, finding a stairwell, differentiating between two nearby bus stops, etc.). Directional RIAS anchored to vehicles and infrastructure (e.g., Talking Signs, 2008), have proven useful in terminals and as stand-ins for audible stop annunciators on buses (Golledge, Marston, & Costanzo, 1998). More detail on the strengths and weakness of RIAS can be found in Petrella, Rainville, and Spiller (2009) and Miller (2012). The main weakness for such systems is their cost (Miller, 2012). This technology is much more expensive to deploy than passive tag approaches (e.g., RFID, NFC, QR Code) or smartphone localization. Thus it is unlikely that this technology will become a mainstream product. Therefore, RIAS is best used in targeted ways as a complimentary technology to benefit individuals who need enhanced information.

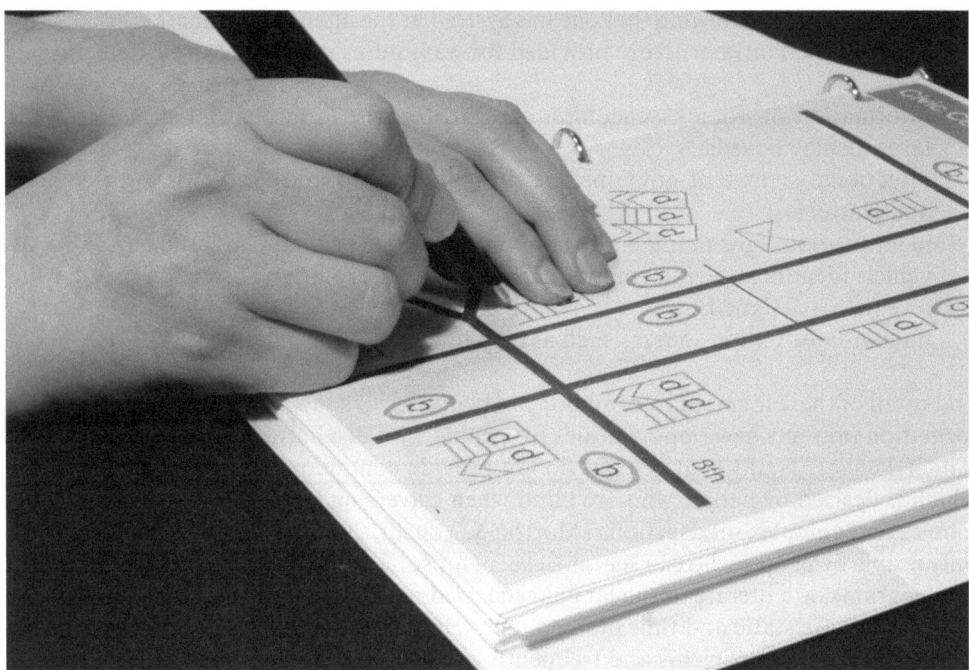

Figure 9.1 Smartpen used with tactile map.
Source: Photo courtesy of Lighthouse for the Blind San Francisco. Photographer: Xander Hudson

Another novel approach to providing location information is the use of a smartpen to integrate spoken labels and guidance with a tactile map (Kehret, Miele, & Landau, 2011; Miele & Landau, 2010). This technology (see Figure 9.1) is powerful and allows rich interaction with the tactile map. Smartpens can cost as much as a smartphone but they do not require monthly data plans and are generally unappealing to potential thieves. The cost of the smartpen technology is much lower than RIAS and this is an attractive option where it is desirable to provide special assistance for visually impaired travelers, for example, in a particularly complex transit terminal. To implement, this technology requires the preparation of annotated plans of each station and area on special paper.

Both these technologies are best suited for stations with dense demand. It is important to note that those who can benefit from their use must also know that the technologies are implemented in the stations they use, obtain training in their use, and have the resources to purchase the equipment. Assuming these conditions are met, the cost-benefit ratio for deploying both technologies becomes problematic for simple bus stops and low-volume stations. The obvious solution in the latter locations is to provide localized information using ubiquitous mobile phone technology.

Inferring Location

Localization is a process in which an individual obtains data from known locations and uses it to estimate their location. For example, a pedestrian in a city reads street signs, sees landmarks, hears the sound of busy traffic and then estimates his or her location by referring to a "cognitive map" or mental image of the area. Navigational software uses a similar process by obtaining information from digital sources in the vicinity and estimating location by referring to a "map" represented by a database of know locations.

Most smartphones offer embedded localization capability through a mixture of cell towers, GPS, and Wi-Fi access point maps. Cell tower localization, which is present even in regular mobile phones and mandated by E911 standards, is the least accurate of the three. It is adequate for determining neighborhood-level location, but not good enough to pinpoint a particular intersection or building. The next level of accuracy is to use a GPS chip to further refine location down to a block. Unfortunately, the satellite-based GPS approach is susceptible to errors in urban "canyons" and does not work well when indoors or under dense tree canopy or metal roofs (e.g., bus shelters). Fortunately, accuracy can be improved dramatically using Wi-Fi access point information. Each location in a city or suburb has a number of fixed Wi-Fi hotspots. A cellphone app can use a map of these hotspots to triangulate a precise location by sensing the Wi-Fi hotspots in the immediate vicinity. This approach even works indoors since Wi-Fi signals are not blocked by the built infrastructure. Triangulation of Bluetooth low-energy tags is also possible (e.g., NavCog, 2015). However, these approaches can be energy intensive and persistent use of high accuracy localization can rapidly drain batteries.

Extensive research has been undertaken to reduce battery consumption for location-based mobile applications. For example Constandache, Gaonkar, Sayler, Choudhury, and Cox (2009) implemented a framework and prediction-based heuristics that improve location accuracy given a certain energy budget. Zhuang, Kim, and Singh (2010) also implemented a middleware framework on Android-based smartphones that piggybacks or adapts applications' location-sensing requests to save energy of GPS use. The SensLoc (Kim, Kim, Estrin, & Srivastava, 2010) system applies a place detection algorithm to reduce battery consumption when detecting place visits and tracking the total travel distance. VTrack (Thiagarajan, Ravindranath, LaCurts, Madden, Balakrishnan, Toledo, & Eriksson, 2009) estimates travel time by using a hidden Markov model (HMM)-based map matching scheme. The Caché (Amini, Lindqvist, Hong, Lin, Toch, & Sadeh, 2011) system pre-fetches potentially useful location-enhanced content in advance to save energy. However, good performance is often achieved by tailoring the approach to the application domain. Lin, Kansal, Lymberopoulos, and Zhao (2010) observed that the required location accuracy varies with location, and dynamically determined the required accuracy for mobile search-based applications. This is a key insight for the transit domain since there are aspects of service delivery that allow software developers to use simplifying assumptions, e.g., buses have predefined routes and stops that limit the potential locations of riders when entering or exiting a vehicle.

It also is possible to leverage infrastructure to pinpoint location without any of these three methods. While not extensively tested or deployed, an RFID, NFC, or QR code tag can be assigned to a specific location and the software can look up the user's position when the tag is scanned with a phone. The installer need only document the tag's physical location in a database. Each tag acts as a pointer to a unique record in the database, providing a low-cost method to identify the user's location. Passive RFID and NFC tags are short-range wireless communication devices similar to transit smartcards. They are extremely cheap, small, easily weatherproofed, require no batteries, and software tools for use with databases are widely available. These features allow novel use cases beyond custom Braille signs at every bus stop. Furthermore, the technology can be used without making contact with the sign, thereby eliminating the need to remove a glove in winter or clear snow and ice off a surface. QR Codes are computer-readable images scanned with a camera phone that point to a specific website address. However, these are easily obscured with snow and dirt, can fade over time, and easily damaged by vandalism. Riders with visual impairments also have difficulty scanning a QR code.

All of these tag technologies present some security problems because they can become vectors for cyber-tampering. For example, one attack was able to wipe data from certain phones through QR Codes or NFC tags (Musil, 2012). Of these, QR Codes are the most susceptible to tampering since they can be masked and replaced by a simple sticker. This easy attack vector, combined with issues associated with weather and damage, suggest implementers should focus on NFC and RFID solutions. NFC tags are much less expensive than RFID tags and can be read by many modern phones. NFC is becoming the standard for phone-based payment in transit settings and a number of phones have built-in NFC (e.g., most new Apple and Android phones). It is quite reasonable to envision a future where (a) a blind user swipes their phone over an NFC or RFID tag at a bus stop and is told the stop's name and which bus will be arriving soon and then (b) swipes again when boarding the bus and told payment has been accepted and their balance. Payment with NFC already exists in some U.S. cities and a precursor RFID system for the blind that links real-time arrival information and when to get off is already on the market (PAVIP, 2009).

Unfortunately, many wayfinding, payment, transit information, and trip-planning systems are not well integrated. One possible future for digital information is the simultaneous operation of many different systems, none of which address all the problems. Without proper integration, the result could be a very confusing set of applications that would challenge all riders, even those without disabilities. For example, imagine a situation where a rider needs to use different smartphone apps for planning trips, purchasing tickets, getting next bus or train information, identifying when one's stop is coming, and finding the right exit from a station. These are just the basic elements of the potential information that a user may need to access in the entire Travel Chain. Thus, a key challenge in implementing a universally designed digital information system is to integrate the various information systems to provide a seamless user experience.

Machine Learning

When a user first accesses an information system, they often are asked to identify their preferences for frequently used information. This is time consuming and a barrier to use of the system. This upfront effort can be reduced through machine learning. This technology can also help the system become more robust over time. It can automatically adjust to changes and will be relatively immune to inaccurate entry of preference data which, in our experience with in-car navigation systems, is quite common. It also will reduce the effort required to enter and access information, a key issue for most people with disabilities.

In machine learning, a computer program *analyzes* the input to a device using a learning model. The model *predicts* the likely next steps in the interaction. The system then *modifies the interaction* to simplify the use of the interface. Finally, the system *observes* the user interaction and uses its observations to improve the learned model. Note that this style of interaction is, in general, very difficult to get right (Faulring et al., 2010; Freed et al., 2008). However, it can be effective in narrow problem spaces. Amazon, Google, and other successful commercial systems already employ this kind of machine learning personalized prioritization. Companies use these techniques to tailor advertisements, auto-complete search queries, and pre-fill web form fields. A simple version of such a system is one operating in Barcelona on the commuter rail network. The system remembers the last ticket purchase associated with a user account and displays that for the first choice. Since the pattern of commuting is often stable, this simple system saves riders significant effort in purchasing tickets. If a trip will be to a different destination, the rider has the option of changing it.

We believe successful implementation of machine learning will produce a major advance in the usability of transit information systems. Data entry into tiny smartphone interfaces is a major challenge for users who are mobile. Location and activity-driven machine learning can pre-fill and auto-complete these steps, thereby driving down data input effort. If a commuter always rides the same routes, then the problem becomes very narrow and straightforward. However, many riders do not have simple trip patterns. Our preliminary analysis of Tiramisu data (Chapter 10) suggests there are opportunities to successfully apply machine learning to more than just the simplest, narrowest cases.

Machine learning can also be used to address the aforementioned system integration problem. Traditional approaches for linking systems together, especially for legacy systems, involve allocating a programmer to write custom software to convert data from one system to another. This can be expensive, time consuming, potentially buggy, and lead to complexities in making changes in the data structure. In many cases, these issues present an insurmountable barrier to implementation. Often, the desired information can be accessed and imported on a periodic basis (e.g., once an hour, day, week, etc.). When real-time links are not needed it becomes possible to "scrape" information off websites using machine learning techniques. A promising approach explored by our team centers on a class of machine learning called programming-by-demonstration (PBD) (Gardiner, Tomasic, Zimmerman, Aziz, & Rivard, 2011). PBD is attractive for use by non-programmers who are interested in extracting the data for novel applications. PBD eliminates the need to maintain specialists in legacy data structures and allows a regular administrative assistant to teach the system which data on a website is valuable and how to navigate the website to find the relevant data. Once taught, the system can mimic these steps to scrape the desired data as needed. Furthermore, the system can identify when the PBD sequence has encountered a problem and needs retraining, thereby mitigating problems with system and data type changes. Initial studies with administrative assistants showed high performance and end user acceptance (Gardiner et al., 2011).

Integration of Geolocation Data with Information about Accessible Infrastructure

In the summer of 2010, the White House initiated a 90-day challenge to pursue and examine how transportation data and other geo-data can be used to increase accessible travel by people with disabilities. A team was convened to explore this issue in depth and report back to the Office of Science and Technology Policy (Geo-Access Challenge Team, 2011). Members of the team included representatives from government, academia, industry, and non-profit advocacy groups.

An overarching theme was the importance of integrated information; riders want the ability to link points of interest, travel data, and municipal infrastructure (e.g., curb cuts, audible crosswalks, etc.) in order to develop a full picture of what they might expect when executing a trip. This type of data is also relevant to transit agencies since it may help determine whether a conditional rider is offered subsidized paratransit since one qualification for paratransit eligibility is the presence of environmental barriers in the path of travel to the closest transit stop.

An excellent example of integrated accessibility information for transportation and points of interest is IBM's AccessMyNYC demonstration (AccessMyNYC, 2012). This smartphone-friendly website allowed users in New York City to find and plan transit or walking routes, identify accessibility information about points-of-interest, and then share their views using ratings and Twitter. While this was a short-term demonstration, it showed the power of this type of integration. There are many examples of less integrated approaches focused on just the accessibility of points-of-interest, stores, and other facilities (e.g., IBM Accessibility City Tag, Wheelmap, CitiRoller, OpenStreetMap, etc.).

While some of the recommendations from the Geo-Access team were specific to real-time information, most were specific to trip planning and information resources for developers. The recommendations were (Geo-Access Challenge Team, 2011, pp. 1–2):

- *User Needs Research*—Produce an annotated bibliography from existing sources of user needs covering the full range of accessible public/private transportation and municipal points-of-interest, and pursue additional research studies where necessary.
- *Information Ecosystem and Business Models Research*—Pursue research studies on how accessible public/private transportation and municipal Point-of-Interest (POI) information is created, collected, aggregated, integrated, and utilized by authorities and citizens/consumers. Also, leverage local public-private partnerships experienced in this to research the various business models that enable cities and regions to offer location-based information and services.
- *Policy*—An institutional and policy assessment—including ramifications related to information security and privacy—should be conducted to include three kinds of data required to enable transformation: Transportation Data (including Accessibility), Municipal Infrastructure Data, and POI Data. The Geo-Access Challenge Team recommends following up the assessment with a Federal Role paper that defines the rules of engagement between different agencies, open data guidance and an information security and privacy white paper to govern these initiatives.
- *Standards*—Enhance existing standards or develop new ones to support structured data collection, aggregation, exchange, and interoperability for accessible transportation, relevant municipal infrastructure, and municipal POIs, to support innovations in location-based information and services.
- *Data Environment Development*—The data environment to support structured data collection, aggregation, exchange, and interoperability for accessible transportation, relevant municipal infrastructure, and municipal POIs needs to be developed, tested, and refined.
- *Technical & Applied Research*—Once policy, standards, and data environment are developed, technical and applied research needs to be encouraged and supported to enable development of innovative applications and solutions. A state of the practice and innovation scan should be undertaken, and technical demonstrations and near-term/long-term development of applications should be supported.
- *Technology Transfer and Implementation Support*—Novel approaches for transferring the enhanced geo-data policies, standards, and data environments into wider usage in both public and private sector arenas should be supported.

The team observed there is a lack of extensive research findings related to how emerging technologies (e.g., social networking, etc.) intersect with accessible travel. This, combined with findings that many apps and smartphone-friendly websites are not accessible (NCAM, 2012) and the general accessibility problems associated with Web 2.0 and smartphone software, suggest that a digital divide is developing between those who need accessible interfaces and those who do not.

Problems

- Precise localization, which means finding one's position in the world, is a major challenge with wayfinding in transit settings.
- Interoperability between wayfinding, payment, transit information, and trip-planning systems is limited, which leads to complex, multi-system use cases.
- Lack of interoperability in backend databases makes it difficult to provide streamlined experiences and location-specific information.
- Information about transit, street infrastructure, and local POIs are frequently hard to obtain due to limited use of standards and restrictive policies.
- Knowing what information is needed when and where for a specific user is challenging.

Recommendations

- Explore methods for providing precise positioning in smartphone apps, even while indoors.
- Use machine learning and data mining techniques to provide interoperability without custom software development.
- Develop methods for tailoring information to the user at the right time and place.

Directions for Future Work

- More advances are needed on ways to provide mobile, tailored, location-appropriate information.
- Technological, standards, and policy advances are needed to facilitate data integration across transit, street infrastructure, and local POIs.
- Advances in these areas should be deployed and evaluated in order to identify and develop best practices.

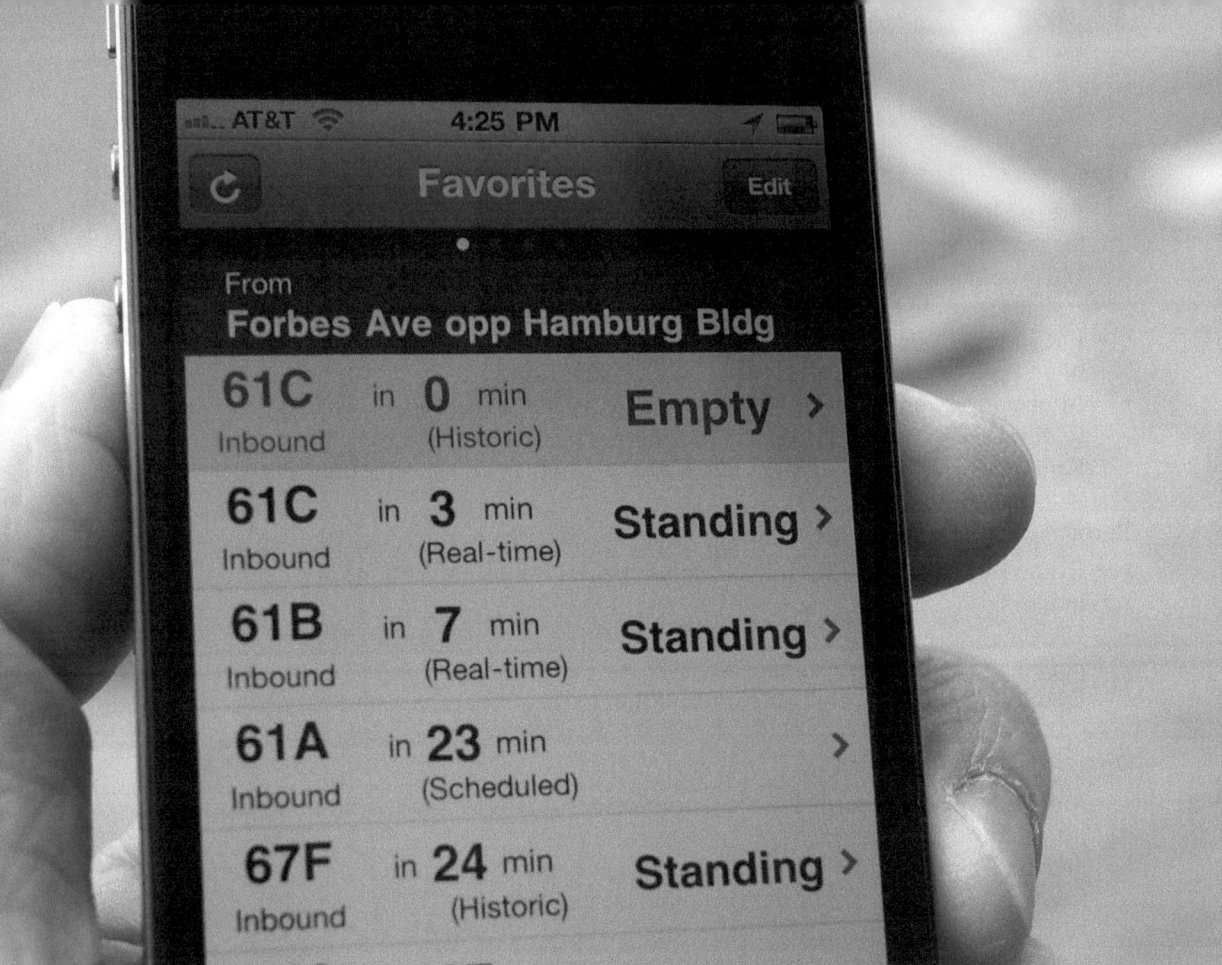

Photo of a smartphone screen displaying real-time transit information.
Source: Aaron Steinfeld

10 Social Computing and Service Design

John Zimmerman, Aaron Steinfeld, and Anthony Tomasic

Overview

Advances in two new fields of practice related to information technology are starting to have an impact in the field of accessible public transportation—social computing and service design. The impact is most obvious in human-computer interaction and transportation service delivery. This chapter describes these two fields, the main threads of development work, and their impact on accessible public transportation. It identifies challenges that need to be addressed to make further advances in practice and avenues for future research and product development.

Social Computing and Transit

Social computing is a large research area that investigates both how the ever-increasing number of computational systems and communication technologies have influenced people's behavior and how new technical communication systems can be designed to engender specific behaviors. Recently, this research community has become interested in how people and computing systems can be combined to form socio-technical systems that can do things neither computers or people can do independently. This focus on combining human and computer resources together has been expressed in the literature in several different ways including:

- *Citizen Science*: people work as sensors and relay information to scientists (Silvertown, 2009). Examples include the Audubon Christmas Day bird count that has been underway since 1900.
- *Crowdsourcing*: combines outsourcing with the concept of the "wisdom of crowds," the idea that many people working a little bit are smarter than a few people working a lot (Howe, 2006). A great example of this can be seen in Wikipedia.
- *Human Computation*: views people as a type of algorithm that can provide information computational systems need (Von Ahn, Maurer, McMillen, Abraham, & Blum, 2008). These systems are often designed as games, where people's playing provides valuable inputs to computational learning systems.
- *Participatory Sensing*: views the ever-increasing number of mobile phones in the world as a new type of instrumentation (Burke et al., 2006).

An example of an application of social computing in public transportation is the use of Twitter by some transit agencies to push just-in-time information out to riders, such as unexpected closures and detours. The Port Authority of Allegheny County (PAAC) is an interesting case study due to two major disruptive events in Pittsburgh—the 2009 G20 meeting and a week of major snowstorms in February 2010. PAAC used Twitter heavily during the crisis, often copying and pasting internal messages straight into the account (Schwartzel, 2010). A byproduct of this decision was a two-way flow of information, where riders also sent messages to PAAC through Twitter about what they saw in their neighborhood. These events led to significant increases in followers (Steinfeld, Rao, Tran, Zimmerman, & Tomasic, 2012) and cemented rider perception of value. Twitter has also been suggested as a source for gathering real-time data from populations during evacuations and public planning (see for example, Brabham, Sanchez, & Bartholomew, 2010; Turner et al., 2010).

Social interaction within transit information systems is also becoming increasingly common. There are several smartphone apps on the market that allow rider-rider dialog and some research groups, including ours, have explored this approach. Work by Machado, Jose, and Moreira (2012) suggested that entertainment and content sharing have potential value, especially as a means for reducing "psychological travel time." Our own work suggests rider-rider dialog and location and route specific alerts from the agency have the potential to increase transit appeal within riders (Steinfeld, Rao, et al., 2012).

Social Computing and Government

The increase of broadband Internet connections in people's homes, the meteoric adoption of Internet connected smartphones, and the rise of social computing, particularly social media applications, have forced government agencies to reconsider how to engage with and communicate with the public. The example of transit services providing up-to-date information via Twitter is one example of this new behavior.

Urban planning has a long history of considering the needs of citizens in the creation of new urban environments (Brabham et al., 2010). Planners most often use surveys, focus groups, and community meetings to achieve some measure of citizen participation, and they note that these traditional methods can be slow, expensive, and often engage only a tiny fraction of citizens. Recently, they have begun to investigate the use of social computing to improve the efficiency and effectiveness of citizen participation (Brabham et al., 2010).

The Urban Mediator offers an interesting example of citizen participation in digital planning (Saad-Sulonen & Cabrera, 2008; Saad-Sulonen, Botero, & Kuutti, 2012). This online tool allows citizens and urban planners to create and share topics. It has been used in Finland to engage citizens in discussing traffic issues and to share their ideas about design proposals. Research on Urban Mediator is particularly interesting because it provides evidence that citizens can and will engage in planning discourse related to their urban services.

There have been several third-party services designed to connect citizens and government services. For example, FixMyStreet.com (King & Brown, 2007) is a service that allows people to report problems like potholes to municipal authorities. ParkScan is a service that allows San Francisco residents to report on problems with parks. ParkScan has a feedback loop that allows users to see how the problems they report have been addressed. In addition, it helps to track the relationship between funding and services.

A recent paper investigating FixMyStreet (King & Brown, 2007) proposes a three-stage framework for effective dialog between citizens and government services: Stage 1, electronic communication between citizens and services; Stage 2, analysis of electronic communications to gain insight on services; and, Stage 3, citizens share information with other citizens and with the service. The authors claim that FixMyStreet's current design cannot reach Stage 3 because citizens post individually, and cannot unite their voices.

eGovernance is a growing research and practice area. This community has recently expanded its focus from providing information and delivering bill-pay and other services over the Internet, to utilizing new communication technology to transform government toward efficient, effective, and citizen-centric service delivery (Qian, 2010). A key component of this transformation is "public consultation": directly engaging citizens in governance by notifying them about government decisions, soliciting public opinion, and actively involving the public in planning and policy formation (Rodrigo & Amo, 2006). Public consultation can take a variety of forms, including petitions, voting, and public comment/discussion systems (Petrik, 2009). Regardless of the mechanism, public consultation is generally regarded as a means to broaden public participation in governance, and increase government transparency and efficiency (Matheus & Ribeiro, 2009a).

Prior studies of public consultation systems have found a direct impact on top-level decisions (Kaschesky & Riedl, 2009). However, it is also acknowledged that many online deliberation and public consultation efforts have fallen short of expectations (for example, Chango, 2007). Among the primary reasons such systems have failed to deliver has been an overreliance on technical and organizational imperatives, with insufficient involvement of the public in design and implementation (Davies, Janowski, Ojo, & Shukla, 2007). Other challenges that public consultation systems face include processing and summarizing large numbers of contributions (Kourmpanis & Peristeras, 2010), encouraging public trust and confidence in government (Tomkins, Pytlik Zillig, Herian, Abdel-Monem, & Hamm, 2010), employing transparent decision-making processes that enable citizens to understand the impact of their participation (Cho & Chun, 2011), and ensuring universal access and quality of information (Matheus & Ribeiro, 2009b).

Service Design

Service design emerged as a distinct design discipline in reaction to the transition in many post-industrial countries from manufacturing and selling products to delivering services. Today approximately two-thirds of the economy for "high-income countries" like the U.S. and most of Europe is in services. Service designers work to conceive of a *service concept*, a description of *what* a service is and *how* it both meets the needs of customers and fulfills a service provider's strategic initiative (Goldstein, Johnston, Duffy, & Rao, 2002). It is important to note that today, almost all technology-driven products and services follow a *product-centric* design approach modeled on user-centered design, but this is beginning to change as the HCI community increasingly considers service design as new approach to conceiving of and innovating services (Forlizzi & Zimmerman, 2013).

Service designers focus on finding opportunities for co-production of value between customers and service providers (Edvardsson, Gustafsson, & Roos, 2005). Services are distinctly different from products in terms of value. In a product-centric view, value is exchanged at the time of purchase; ownership is transferred from the maker to the consumer. From a service perspective, value emerges during a customer's interaction with a service provider. There is no exchange of ownership; instead value is co-produced between the service provider and the customer. This idea of co-production of value emerged partially out of the insight that customers can be more than consumers; they can also be a source of competence (Prahalad & Ramaswamy, 2000). As an example, Prahalad and Ramaswamy note that in the software industry, customers commonly share their competence by becoming beta-testers (Prahalad & Ramaswamy, 2000). This insight on customer competence as a critical resource arose from the Internet's ability for companies and customers to engage in dialog, and also from the recognition that dialog between the customer and the service provider as well as within communities of customers presents an opportunity for new kinds of value to emerge (Prahalad & Ramaswamy, 2000).

The service design community has noted that public services are a special case for service innovation. Often these services function as monopolies and are not driven to improve their service offerings through competition. This fact is not necessarily a negative quality, as competition can diminish the importance of marginalized communities like the poor and people with disabilities, who are often the intended audience for public services, and in fact, many public services exist specifically to address issues of fairness by creating a more level playing field for marginalized communities (Boyne, 2003).

Public services, similar to eGovernance, have looked to technology as an approach to service improvement. Traditionally they have focused on automation and on improving the speed of service delivery. However, when public services take a more user-centered design approach, they often learn that their customers' main desires are for different services, not for automated services (Bradwell & Marr, 2008). In detailing opportunities for public service improvement, researchers have noted seven metrics to assessing service improvement/innovation:

- *Quantity of outputs*: More services offered.
- *Quality of outputs*: Speed and reliability.
- *Efficiency*: More outputs for less money.
- *Equity*: General sense of fairness toward citizens in terms of costs and benefits.
- *Outcomes*: Increased usage of a service by the citizens or by a percentage of the citizens.
- *Value*: Cost per unit of outcome.
- *Consumer satisfaction*: Often a result of several of metrics above (Boyne, 2003).

Researchers have proposed that co-design, where customers engage in the design of the services with the provider, may be a particularly good approach for service improvement of public services. They noted three distinct beneficial outcomes: (1) services that are more responsive to the changing needs of citizens; (2) increased trust of the government through citizens more positive engagement with services; and, (3) building of social capital through an increased sense of community (Bradwell & Marr, 2008).

Case Study: Tiramisu

Our team developed a social computing transit rider information system, called Tiramisu, to act as a testbed for research and demonstrate how universal design, social computing, and service design can be combined to provide value for all transit riders (Steinfeld, Zimmerman, Tomasic, Yoo, & Aziz, 2012; Zimmerman et al., 2011). The Tiramisu system allows riders to generate and share GPS traces and fullness (whether there are many, some, or no seats left on a bus) ratings via their mobile phones. This data is then aggregated in order to provide real-time arrival and fullness information. The riders collectively generate the information they desire, circumnavigating the need to install expensive commercial systems. In addition, Tiramisu allows riders to report problems with the service. From the perspective of accessible transportation, Tiramisu provides an accessible social computing tool that can give valuable information to riders with disabilities at no cost, improving usability of transit systems. It also provides a vehicle through which riders with disabilities can participate in improving service delivery by communicating their perspectives to the operating agency.

Social computing systems and other online communities must deal with the issue of motivating participation. While many crowdsource models benefit from altruistic participants and a sense of ownership (e.g., ParkScan.org), our evidence suggests that this behavior is not that common among transit riders (Yoo et al., 2010). We suspect this difference is based on two factors. First, riders view transit as a "means" rather than the "ends." In a sense, riders engage with transit service not for the specific experience of the ride, but in order to efficiently achieve a different goal that requires them to move within a city. Second, riders of public transit services interact with the service much more like a consumer than for other public services (e.g., visitors to public parks). Riders pay a small amount for each journey or they repeatedly purchase passes. This constant financial transaction may frame riders' perception of public transportation as a consumable service as opposed to a public good enabled by taxpaying citizens.

Therefore, systems focused on transit need to provide value to the user rather than simply rely on altruistic and ownership-induced behavior. Real-time information about arrival time and vehicle fullness were identified as high-value items by riders with and without disabilities. Furthermore, we focused the appearance and language of Tiramisu to embody universal design principles so it would appeal to a wide audience. Wide appeal is critical in social computing systems since performance usually improves as the user base grows.

Tiramisu allowed users to find nearby bus stops through a map or list, as shown in Figure 10.1 (a, Main Map). The system provided predictions of arrival times based on scheduled, historical, and real-time location data and ratings for bus load ("No Seats" in b). Historic data was generated from prior rider contributions and offered when no user was contributing in real-time. Users identified the bus they are boarding, their destination, and rate the bus load (c) and shared a location recording while riding (d). When users wished to file a report, they first selected whether the report was specific to the Tiramisu system, the schedule, and predictions, or the local transit agency. For the latter, they were offered high-level categories (e) prior to the actual report screen (f). On the

Figure 10.1 Tiramisu interface screens. (a) Main map, (b) select route, (c) before recording, (d) during recording, (e) agency report categories, (f) filing a report.

Source: Aaron Steinfeld

report screen, tags were listed and the text field is pre-populated with location and route information so users would realize they do not need to provide such details. Users were also given the ability to include a picture. Text and pictures were identified in our earlier work as the preferred and most effective modalities for reporting accessibility barriers in transit scenes (Steinfeld et al., 2010a, 2010b). Additional detail on the design process, system architecture, and rationale for various features can be found in earlier papers (Steinfeld, Zimmerman, et al., 2012; Yoo et al., 2010; Zimmerman et al., 2011).

An early version of the system was deployed and pilot tested as a closed beta (Zimmerman et al., 2011). The goal for this pilot was to find software bugs and improve algorithms, so riders with disabilities were not explicitly recruited and none participated. Since accessibility was a critical goal even this preliminary version fully supported screen reading using the iPhone's VoiceOver feature. The publicly deployed iPhone version is also VoiceOver compatible.

The pilot included data from 28 people over a span of 38 days. Participants returned after a few weeks to complete a survey and receive payment for their assistance. Riders were recruited from a specific region of the city to increase the chance their contributed data would directly benefit their fellow participants. This appeared to be the case and sampling of stops within this transit corridor frequently revealed at least one real-time and several historic buses. Rider data from a 21-day period was used for each round of analysis, with a focus on the 14 days prior to payment and seven days after payment. Riders contributed data for 56 percent of the time when using the app. Individual rows in viewed arrival pages (e.g., 61C inbound in five minutes) shown to participants contained either historic or real-time data 13 percent of the time (2,132 out of 16,263). This is a very good rate, given the number of possible buses and the limited number of participants.

Participants in the pilot predominately reported issues with the system, which was expected and desired since this was a software beta test. However, 14 participants submitted a total of 22 reports specific to the transit system. In prior work (Steinfeld et al., 2010b), riders who use wheeled mobility devices reported at a much higher rate (58 percent of respondents) than their peers without disabilities (17 percent). The latter aligns well with the data seen in this pilot.

As of this writing, the system has now been in public operation in Pittsburgh, Pennsylvania, since late July 2011 and users recorded over 210,000 contributions by the winter of 2017 and use the app more than 1.2 million user-days. The iPhone version was released at launch while the Android version followed in the fall of 2011. The system has also been deployed to New York City. Analysis of the data suggests our approach works, especially during rush hour periods (Tomasic et al., 2015). It is worth noting that the vehicle fullness and rider origin-destination pair data collected by Tiramisu would be extremely difficult to collect accurately using traditional transportation information systems. For example, most transit agencies only know how many people entered or left a bus—they have no way to link the origin and destinations of riders. This data has significant value for route planning and daily operations.

The first 1,000 reports filed in Pittsburgh were analyzed for behavioral trends, emergent themes, and the implications of these findings (Steinfeld, Rao, et al., 2012). As in earlier research (Zimmerman et al., 2011), riders provided transit agency feedback reports through the system at rates much higher than would normally be expected. There was strong representation for reports related to "point-of-pain" experiences, often related to missing or late buses. Other comments included driver feedback, maintenance issues, and thoughts about fellow riders. These reports suggest a willingness of users to report from bus stops and other transit environments in real time. As expected, the system captured instances in which drivers would not let wheelchair riders onto the bus. This confirms laboratory results that implied riders would use mobile reporting to document such events (Steinfeld et al., 2010b).

One unforeseen aspect is the interaction between user-contributed data and existing service quality policies. For example, third-party reporting systems, like Tiramisu, may not be considered acceptable documentation for legal or disciplinary action under existing contracts. Route planning, non-contractual service modifications, and other co-production efforts are less likely to encounter resistance to third-party data contribution. Therefore, we recommend focusing on rider-to-rider communication for peer support, alerts from transit agencies, and methods for creating dialog on issues not tied to disciplinary action. Transit agencies interested in mobile reporting of data requiring traceable provenance, like disciplinary problems, should develop mobile versions of their existing website complaint forms that are friendly with VoiceOver (iPhone), TalkBalk (Android), and other smartphone screen readers. Finally, these agencies should also develop protocols for using third-party data to help triangulate acceptable data within their own systems.

Problems

• There is increasing interest and demand to establish citizen-provider interaction through social computing models.
• The fields of social computing and service design are still rather young and have yet to be heavily deployed in disability and transportation domains, thereby limiting opportunities to learn from best practices.
• As with most new technologies, there are mismatches between social computing solutions and traditional public entity processes.

Recommendations

• Explore the use of social computing (e.g., Twitter, Facebook, etc.) in public transit, especially for rider-contributed data on system status and condition.
• Dedicated adequate staffing for social computing and use supportive policies for two-way dialog.
• Consider creative uses of social computing and service design to serve user needs while evaluating and implementing more expensive, comprehensive alternatives.
• Engage end users to provide value and content to service offerings.
• Utilize universal design to foster input on accessibility issues from people without disabilities. For example, riders of all abilities appreciate not having to step into the street when boarding a bus.

Directions for Future Work

• There are more opportunities to combine social computing, service design, and universal design.
• More innovation is needed on how to utilize social computing and service design to support independent mobility around the community.

Photo of an older woman highlighting a route using a map.
Source: Aaron Steinfeld

11 Learning from Riders

Jordana L. Maisel, Daisy Yoo, John Zimmerman,
Edward Steinfeld, and Aaron Steinfeld

Overview

If accessible public transportation is viewed as a service design problem, the effective-ness of a system should be measured by the experience of the riders. This new approach takes a consumer perspective in which the goal is to address the riders' needs effectively rather than just improve a system's level of compliance with design standards and rules. There are significant problems evaluating service delivery since experiences can change from place to place, day to day, and even minute to minute. This variability presents a difficult problem in evaluation, but new ideas and technologies offer some promising solutions. Some methods for capturing knowledge from riders include guided tours, real-time reporting, online surveys, and diaries. A multi-method approach to evaluation is warranted since each method used alone will always have limitations.

The term "services," in the context of accessible transportation is generally meant to include traditional operational issues like:

- on time operation;
- maintaining equipment facilities such as lifts and ramps on buses, fare machines and gates, and elevators in stations;
- customer services like fare collection and assistance boarding.

A more comprehensive definition of services extends beyond operations to include many other aspects of service delivery, for example:

- route planning that supports accessibility to service and reasonable ride lengths;
- vehicle selection or design to serve rider needs by providing safe and comfortable access and seating;
- information services like websites and next stop information on-board vehicles;
- training for drivers and station staff on compliance with policies on how to treat people with different types of disabilities;
- training and information for riders to help them utilize the system effectively;
- maintenance of bus stops and stations;
- security on vehicles and in stations;
- effective fare and purchasing policies;
- ensuring that approaches to stations and stops are accessible, safe, and secure.

Another aspect of service delivery in public transportation is coordination with other public agencies and private organizations including:

- security and emergency response with local public safety services;
- providing arrangements to allow vendors to operate effectively on public transportation properties;
- advertisement policies that provide revenue to support construction of bus shelters and station amenities;
- effective transfers between transportation modes, often operated by different agencies (e.g., commuter rail with local bus transportation).

Riders as Co-Designers

Zimmerman et al. (2011) note that public services like mass transportation are monopolies that lack the motivation for change driven by competition in an open market. Their monopoly status reduces the incentive to make system improvements that benefit end users. Likewise, public agency funding and decision-making structures are often restricted via regulations or configured in ways that apply negative pressure on changes. To address these problems, researchers have proposed that customers of public transit be enlisted as "co-designers" (Yoo et al., 2010; Zimmerman et al., 2011). Co-design can take the form of structured design and planning activities like citizen participation in planning a new transit line. Or, it can take the form of engaging citizens in identifying and addressing problems in existing systems. The research reported here focused on the second and focuses on ways to move from limited and periodic "audits" to a continuous quality improvement process that enlists "user experts" in helping to improve service delivery. Identifying and addressing accessibility problems in existing transit systems is hindered by the size and complexity of transit systems and the limited effort agencies can dedicate to identifying problems given their funding challenges. Likewise, consumers regularly report little to no feedback when submitting reports of problems and lack of information on how to make such reports. Thus, it is important to develop methods that will overcome this "black hole effect" but not be financially burdensome to a transportation agency. The authors have developed and tested four such methods.

One method is called Guided Tours. A trained guide invited participants to complete a route through a predetermined part of a transit system (Steinfeld et al., 2011). The participant rated the level of difficulty they had completing different activities (e.g., boarding a vehicle, hearing driver announcements) as they navigated the route and responded to a set of open-ended questions after each segment of the route. The guide probed to identify the reasons for problems identified. The method was tested in Buffalo, NY, with five groups of participants ($N = 50, n = 10$): manual wheelchair users, people with hearing and communication limitations, visually impaired individuals, people with mild cognitive impairments, and frail elderly individuals. Data collection occurred over a 12-month time span in 2010, which exposed participants to a wide range of weather conditions, an important aspect of accessibility in all but the most benign climates.

Results from the Guided Tours study identified many problems but only a few that are currently covered by U.S. accessibility standards for public transportation. The system was generally compliant with current Federal regulations on accessibility; thus, the Guided Tours methodology is sensitive enough to identify issues beyond the basics of accessible design. Results of the Guided Tours implementation demonstrated that problems with information access and understanding the system were common across most groups of participants (i.e., mobility, visual impairments, frail elderly, cognitive impairments, and hearing impairments) and were the most serious problems identified (see Figure 11.1). This included problems with access and usability of system websites, obtaining and understanding maps and schedules, using ticketing machines, and finding transit stations and stops. Access to the pedestrian environment around stations proved to be the most common problems with mobility but many of these problems were related to perceptual and cognitive issues. The results emphasize the importance of the information and communications environment, design for ease of understanding, and pedestrian rights-of-way as cross-disability accessibility issues (Steinfeld, Grimble et al., 2012).

The Guided Tours methodology proved to be a simple method for doing periodic audits of a system's usability. It identified many problems that would not be obvious during a standard accessibility audit. Furthermore, it was relatively inexpensive to implement. One limitation of the Guided Tours method is that it proved to be unrealistic, given the scope of the study, to meet people at their home in order to study the critical segment of the Travel Chain from home to bus stop. A second limitation is that we could not study more than one route circuit with the resources available. But the same methodology could be used to expand the study incrementally. Ideally, the Guided Tours method would be adopted as a continuous process with funding in an annual budget. Evaluations of different sites/routes could be prioritized and completed over the course of several years. Enlisting advocacy groups and volunteers could help to evaluate more routes each year. Training interviewers is a critical concern for ensuring reliability and avoiding bias in such research. Use of university students in field placement and unpaid internships could be another approach to implementing a Guided Tours program with little or no cost. This would also provide assistance of faculty for training, supervision of data collection, and analysis. Further, it would be a good exercise for training future professionals such as civil engineers, architects, and product designers in the details of accessible transportation.

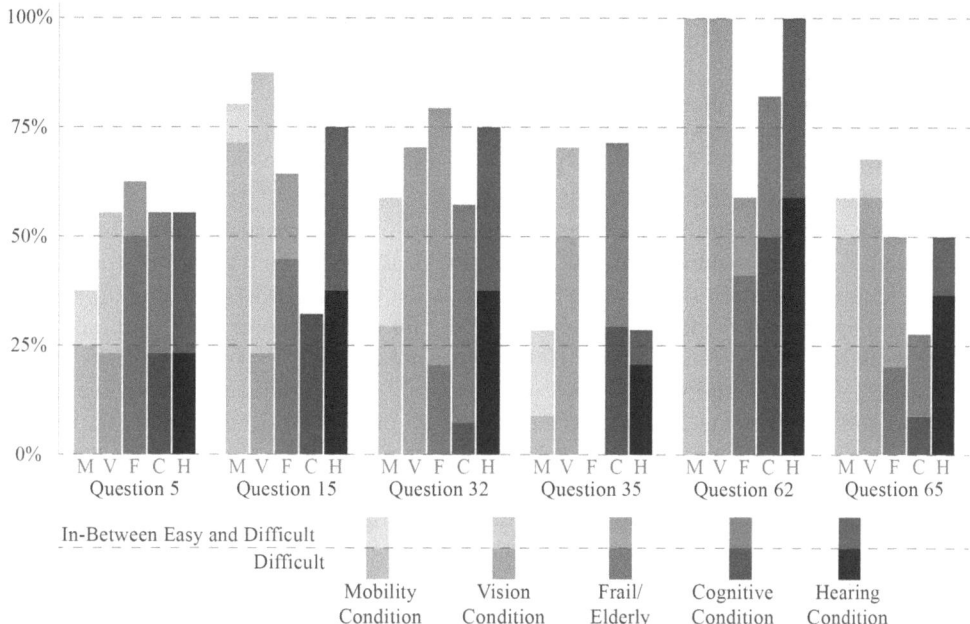

Figure 11.1 Results from the Guided Tours study. Questions Asked Relating to Ease of Task: (bold indicates responses are graphed above). How difficult was it to … **(5) plan your trip**; (8) find information about the routes; (11) understand the information you found about the routes; **(15) understand the schedule and route maps**; **(32) to select the correct ticket to purchase**; **(35) complete the ticket purchase**; (55) get off the train; **(62) use the schedule and route map to find a bus stop**; **(65) find the bus stop**; (68) use the bus stop's shelter; (80) pay the bus fare?

Source: IDeA Center

The second method involved "Real Time Reporting." In this method, transit riders reported on problems they identified as they used the system on a regular basis. To accomplish real-time reporting, we developed a new smartphone application, Tiramisu (see Chapter 10). The daily use element focused on real-time bus arrival and available seats. It also allowed riders to photograph and write a description of problems they encountered along a journey and transmit these reports to a public website. Early development of the system included design and testing with people with disabilities to ensure that it would be accessible and to identify the best modalities for reporting problems (Steinfeld et al., 2010; Yoo et al., 2010). Tiramisu was deployed and available for everyday use and a large archive of responses was amassed, providing new insights into accessibility issues as well as other rider concerns. Problems were posted on a public website and reports relevant to the local transit agency are forwarded regularly. Unfortunately, the agency was unable to use most of these reports due to conflicts with data provenance and internal rules regarding enforcement actions.

Results of the initial testing of the Tiramisu application demonstrated that the number of riders reporting problems increased from 10 percent to 50 percent post-test (Zimmerman et al., 2011). The full-scale field-testing and deployment underway in Pittsburgh is providing extensive information on the types of problems reported and the impact on transit system response. A recent analysis of over 1,000 reports submitted by the general public revealed a variety of problems of relevance to people with disabilities (Steinfeld et al., 2012). For example, there were multiple reports detailing instances where drivers would not let wheelchair riders onto the bus. A meta-analysis of emergent themes revealed both positive and negative feedback regarding (a) schedule performance, (b) bus and bus stop cleanliness and maintenance, and (c) driver behaviors. These themes may seem broad from a disability-centric perspective, but they reflect the overall service quality experienced by riders. A recurring theme was rider desire to communicate with the transit provider. PAAC already had a robust communication approach, including a call center, web forms, a blog, and a very active Twitter account. However, riders still wanted more communication (Steinfeld, Rao, et al., 2012).

The high emphasis put on emotional issues of the transit experience should not be undervalued. It has implications for both general transit service and accessible transit. The factors identified are critical for forming a good opinion of service quality that helps to encourage mode shift from private vehicles, maintain ridership over time, and reduce the desire to abandon fixed-route transit in favor of private vehicles. For people with disabilities specifically, ensuring a good emotional response can reduce reliance on expensive special route or paratransit service and lead to increased support for investing in the accessibility of mass transit in general as opposed to investment in such specialized and expensive services.

The Guided Tours and Real Time Reporting provide two means to incrementally improve the accessibility of existing transit systems by engaging riders in "co-design." Both these methods can be used in an advocacy mode to identify and demonstrate serious inadequacies in systems, but they can be more effectively used in cooperation with the transit agency to systematically identify ways to improve services over time. The Guided Tours method provides an objective "status check" on the accessibility and usability of systems for different groups of riders at a fixed point in time and allows the agency or a third party to focus on specific issues that are critical for decision making; for example, bus design when the agency is planning on buying new vehicles or snow removal at bus stops, to support negotiations to improve service from public works departments or cooperation from local business improvement districts. The Real Time Reporting method allows continuous data collection and helps operators to demonstrate their commitment to continuous quality improvement through a website that documents how problems identified by riders have been solved. Both methods can be implemented independently of the transit system, empowering users and advocates. Both methods conceive of public transit systems as a service, including physical, informational, and human components. By putting the focus on individual experiences rather than a predefined "audit" process, they identify targets for improving the overall quality of transportation services, with a focus on the most important concern of riders.

A third means to engage riders in "co-designing" are consumer surveys. Since the two research activities above were each conducted in one city, they may not reflect priorities in other locations. Thus, the lead author of this chapter led a team that conducted an online survey September 2010–April 2011 to investigate user requirements for patrons of public transit buses. The research studied individuals' experiences: boarding and disembarking public buses; circulating inside buses; understanding the communication and information systems of public buses; and, issues concerning rider safety. The survey specifically sought to answer three research questions:

1   Which features related to using public buses merit future detailed experimental research?
2   What/where do users identify as accessible public transportation best practices?
3   Which features need to be addressed when designing the next generation accessible transit bus?

The survey asked individuals how often specific conditions affect their ability to perform routine activities. The following figure shows the results from the 372 participants (Figure 11.2).

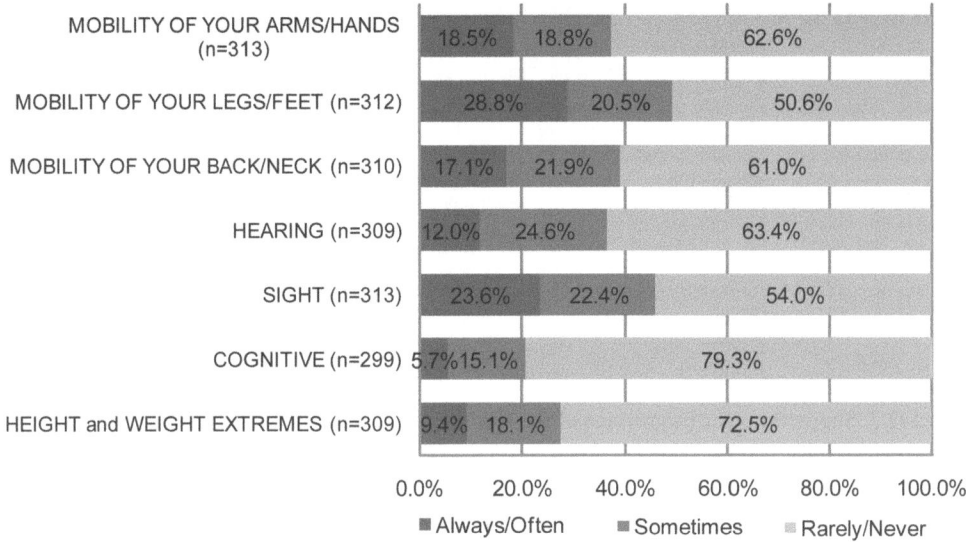

Figure 11.2 Conditions affecting routine activities.
Source: IDeA Center

Participants were then asked a series of questions related to how often they experienced problems with various components of public transit buses. Table 11.1 lists the top ten problems reported, by feature.

The survey also asked individuals to provide "excellent examples" of accessible public transportation. From the 166 respondents to this question, Washington, DC, yielded the most favorable responses with eight percent (*n* = 13). For example, one respondent said, "The DC system has excellent signage; it is well-lit and the cars are spacious." While five percent (*n* = 8) cited the BART and Muni bus and rail system in San Francisco, CA, or the MARTA in Atlanta, GA, four percent (*n* = 7) cited specific features of the New York

Table 11.1 Top ten reported problems experienced by transit bus riders

Total # of Responses	Feature	Responses	% of Responses	# of Responses
n = 316	Announcements	Announcements unclear or muffled	54.4	172
n = 326	Seating	Bus moves before sitting	41.1	134
n = 329	Space Clearance	Blocked aisles	40.1	132
n = 352	Identifying Bus	Route information missing	37.5	132
n = 326	Stairs	Stairs too high from the ground	35.3	115
n = 324	Fare Payment	Payment system confusing	24.4	79
n = 297	Lighting	Bus is too dark	22.9	68
n = 294	Handrails	Handrails difficult to grasp	17.7	52
n = 324	Lift	Lift did not work	17.6	57
n = 309	Securement System	Driver didn't know how to use the securement system	16.8	52

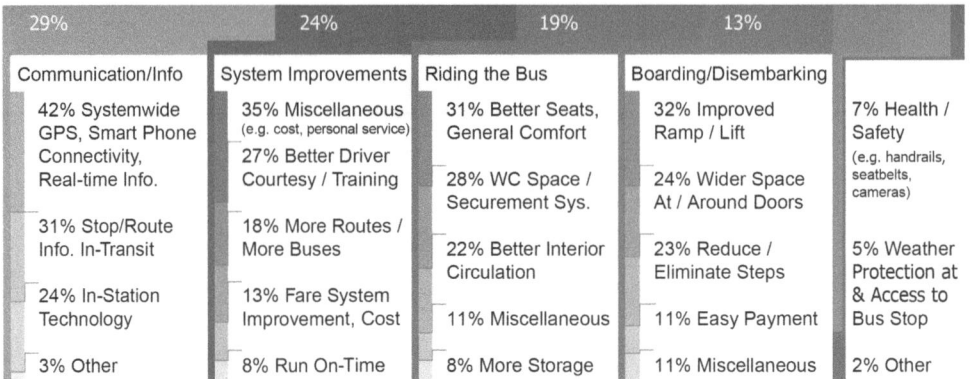

Figure 11.3 Suggestions for bus service of the future.
Source: IDeA Center

transit system or the buses in Chicago, IL. "In the city of Chicago all buses have ramps, no stairs, all have audible announcements, and drivers are courteous." Clearly the results reflect local experience.

Finally, we asked respondents to identify features they would like to see in the next generation of accessible public bus service. The following figure depicts the survey participants' collective vision (*n* = 209) for the suggested improvements and ideas for the future (Figure 11.3).

A final method to engage riders in "co-designing" involves capturing rider data through diaries (Yoo, Zimmerman, & Hirsch, 2013). Members of our team conducted a study with 15 local public transit riders, exploring the issues and controversies over the service they both pay for and use. In particular, we wanted to understand where they wanted to go but could not due to barriers in the transit service, as well as the issues that reveal points of conflict and the groups engaged in this conflict. We wanted to better understand this design space in order to address how information should be structured and displayed within a co-design social computing system. We assumed route planning would be one of the main topics of any co-design system involving transit. Based on this assumption, our fieldwork was framed to elicit issues around the current service offerings and ideas for how routes might be planned differently and preferably in the future.

We chose to focus on riders at a single bus stop, seeing them as a microcosm of the larger set of riders (Yoo et al., 2013). Our study location resides on the edge of a commercial district where it connects with city suburbs. The stop functions as a departure point for people who live in the neighborhood, as a destination for people who work and shop in the commercial district, and as a transfer point for people who live in the suburbs.

We carefully selected participants in order to investigate multiple dimensions of potentially contesting communities. There were six seniors (ages 60–93), eight younger riders (18–27), and one wheelchair user (58). Five of the participants (two young, three senior) lived in the suburbs. The rest lived near the stop and considered it their home stop.

This study used a deep, rich, qualitative data approach. Participants were given a one-week journal consisting of open-ended questions inspired by cultural probes (Gaver, Dunne, & Pacenti, 1999). They were also interviewed in-depth prior to and following the week-long journal period.

We observed that the personal narratives participants shared describing their service experiences could be a rich source of information on how they perceived the actions of the transit service and their fellow riders. A key output from the older and younger participants was the need for communication of rationale and consequences. We identified several areas for potential future work: (1) enhancing dialog with riders about rationales for current service offerings; (2) helping riders share detailed consequences of service changes; (3) helping riders take action and participate in service design; (4) identifying currently ill-defined, yet implicit stakeholder groups; (5) visualizing money flow for better understanding of funding restrictions and sources; and, (6) providing a continuous, traceable history for co-design activity. Knowledge from this study was fed into the larger Tiramisu effort and is informing the design and implementation of the upcoming rider-rider and rider-agency messaging functionality. A detailed paper on this study and the specific rider experiences is currently under peer review.

Many of the problems identified by riders in the Guided Tours, Real Time Reporting, Transit Usability Survey, and Diaries are not covered by accessibility regulations, yet they can present equally serious barriers to the usability of a system. They also illustrate that once the legal requirements of minimum accessibility are met, people with disabilities still face other problems many of which are similar to those faced by other riders and reflect general service design issues rather than specific disability-related issues. Thus improving accessibility, safety, security, and usability in general are just as important as meeting minimum accessibility mandates.

Conceptualizing evaluation from a services perspective rather than a compliance perspective helps to identify the full range of problems faced by transit users with disabilities. Moreover, targeting service quality makes evaluation processes more valuable to the transit agency because they address issues important to all customers. People with disabilities want the same level of service quality as the rest of the population. Although each disability has some unique needs, taken as a whole, people with disabilities can be viewed as a good set of "reference users" to use in evaluation of service quality because they are likely to be more sensitive to problems. Other riders simply adapt with some inconvenience, elect to not complain, and quietly lower their opinion of the transit agency. In particular, using the services approach provides access to emotional responses and cognitive issues in usability that are not addressed in audits of regulatory compliance. Our findings show that these aspects of service delivery are highly important to all transit riders and critical for measuring outcomes.

Problems

- The compliance approach to evaluation of accessibility is too narrow. It ignores issues that are not covered in accessibility regulations but are important to riders.
- The service design paradigm is a new way of thinking about transportation and is not yet adopted extensively in practice.
- Service evaluation is a means to engage transportation users in co-design activities.

Recommendations

- Use social computing to obtain evaluation information at no cost to service providers.
- Use the Guided Tours approach as an easy to implement and inexpensive systematic evaluation technique.
- Use online surveys to verify concerns raised during smaller sample size methods.

Directions for Future Work

- Collaborations with advocacy groups and universities have the potential for helping operators conduct systematic evaluations through Guided Tours. This method is well suited to class projects.
- Service providers need better ways for continuously gathering and summarizing service delivery breakdowns.

Photo of a seat on a train and a view out the window.
Source: IDeA Center

12 Vision for the Future

Aaron Steinfeld and Edward Steinfeld

Overview

In this section, we provide a look ahead to the future of accessible public transportation. As noted at the beginning of this book, the focus of many research efforts has been on intra-city transportation and primarily buses. Thus, we have also focused this book on this sector of the public transportation industry. Much of what is described in this book can be applied to other transportation modes and intercity operations. For example, factors which impact boarding and disembarking of buses apply to most mass transportation vehicles. In this section we identify key challenges for the future in the intercity transportation realm and also for other types of transportation systems. Summarizing and reflecting on the material in the earlier sections, we offer recommendations on priorities for the future related to intercity transportation. Extending the line of work described here to other issues of community participation—such as access to education facilities, health settings, employment, and civic engagement—we offer some speculative ideas that could be fruitful in advancing the field.

Key Challenges

This book contains many detailed challenges associated with specific aspects of accessible public transit based on various activities we completed and our own research. These can be synthesized into the following broad challenges for research and practice:

Seamless boarding and disembarking of vehicles continues to be a significant challenge for many users. Work by the authors has identified specific barriers to rapid ingress and egress from vehicles and has generated some evidence-based design solutions to critical problems. However, there are still many unanswered questions and design strategies that have not been evaluated, especially for rail systems and street cars, since we have focused on the current paradigm for buses, the low-floor bus with a ramp. Our research demonstrates that a systems approach is needed to improve practices in boarding and disembarking. In particular, in addition to ramp and lift design, the location of entries/exits and fare payment systems are important factors in the accessibility and safety of boarding/disembarking. Advanced technologies for boarding and disembarking need to be investigated. The problem of ramp slope is constrained so much by vehicle sizes, for example, that ultimately, more effective solutions to serve all riders may require development of radically different systems.

The pedestrian environment continues to be a source of challenges during boarding and disembarking. The relationship between exterior topography and climate and entry/egress activities needs research. Current regulations do not consider contextual factors, leading to a need for new policy ideas and technology innovation. Developing strategies to address the "last mile" problem is another important issue. The availability of accessible, safe paths between a stop and a destination can be the difference between use of fixed route transit and paratransit. This will require collaborative interventions between transit agencies and other municipal authorities. The relationship between land use and service is extremely important. Transit service can be most easily facilitated by Transit Oriented Development (TOD) like those proposed by Traditional Neighborhood Development advocates. Thus, the research community, transit agencies, and advocates need to become more important players in regional and urban planning activities. In station and terminal design, the main challenges are the identification of best practices and integrating accessibility into information technologies effectively.

Developing strategies to serve low-density communities and neighborhoods is a persistent challenge due to the history of land development and migration since the Second World War. Transportation providers need to find ways to serve people who are dependent on public transportation. One promising approach is cooperative agreements with employers and schools through which the same vehicles can be shared to provide services for commuters and students during peak hours and for recreation, shopping, and health care trips during off-peak hours. Accessible taxis are a strategy that other countries have used effectively to plug gaps in fixed route service and avoid reliance on paratransit. Taxis have the flexibility to be deployed when needed and serve other populations when not. Other forms of on-demand transportation like owner-operated jitney services and TNCs (e.g., Uber), if accessible, may also be effective. For this challenge, strategies are known and in practice elsewhere but they need to be disseminated and adopted more widely so the challenge is more about innovation diffusion than it is about technology.

Implementing accessible information and communication services that are also user friendly are a major challenge for all transportation agencies whether they are urban, suburban, or rural. These services require significant investments to establish and maintain. They help riders to understand the system, complete trip planning, and provide information en route through mobile phones. Progress has been made in the provision of accessible websites. There are two dimensions of website design. First, the information on the website must be accessible to people with visual disabilities including those who use screen readers, those with hearing impairments, and those with dexterity problems that limit their use of interfaces. Second, the content of the websites should provide adequate information on the accessibility of the systems. We completed a pilot study to evaluate the accessibility of the websites of major urban transportation systems. Ten systems were evaluated and tested for accessibility. The degree of accessibility varied significantly and changed rapidly, even from day to day, as updates were added. Updates sometimes improved accessibility and sometimes made it worse. The pilot study showed that accessibility content is highly varied, with some systems like Washington, DC, or the Bay Area Rapid Transit (BART) including much more content and more sophistication in trip-planning tools. As mentioned earlier, smartphone applications are also frequently inaccessible. Automated phone systems are becoming more advanced, but natural dialog interaction appears to only be provided in one city and as a result of a research project (Pittsburgh, PA).

Providing real-time information, as is now expected by riders of all abilities, is an economic challenge for transit agencies. Some of the larger transit systems provide real-time information on broken equipment like elevators and construction or service problems that limit accessibility. Some systems inform riders about detours and alternative routes when there are temporary problems with a station or route. The growing use of smartphones provides an opportunity to support travelers en route, particularly individuals with cognitive impairments, a group that shows heavy use of paratransit. One successful approach, the Travel Assistance Device (Barbeau et al., 2010; Winters et al., 2010) demonstrated this is a feasible goal. The Tiramisu smartphone app, developed by our team (Steinfeld, Rao, et al., 2012; Zimmerman et al., 2011) incorporates crowdsourcing and a social media approach to provide riders with real-time information on wait times and vehicle fullness. We have demonstrated that the general population can be enlisted, using universal design approaches, to generate and contribute information valuable to riders with disabilities.

Service quality continues to be a challenge in public transit. Our research shows that riders desire a rich dialog with their transit agencies about service delivery and quality (Steinfeld, Rao, et al., 2012; Yoo et al., 2013). New social computing techniques like social media and crowdsourcing can provide more rapid and greater visibility and understanding of service quality breakdowns. Such approaches also have the potential to create and foster rider-rider as well as rider-agency collaboration. Co-design through a variety of citizen participation methods offers great potential for improving quality to meet the priorities of transit riders. Co-design can lead to more responsive and appropriate policies and practices. However, application of these new, computer-aided approaches in the transit domain is still underdeveloped and effective interaction models are still being identified.

Location-based information is an important determinant in the last mile. Work by the Geo-Access Challenge Team (2011) highlighted many accessible transportation problems that could be ameliorated with integrated transit and local data. For example, advanced knowledge about the lack of an accessible path to a bus stop due to construction or debris can allow riders and agencies to adapt rapidly. Data interoperability, disparate data repositories, restricted access to public data, and other barriers are limiting innovation in this space.

Rapid and accepted forms of wheelchair securement are still needed for mainline public transit vehicles. Researchers at the now defunct RERC on Wheelchair Transportation Safety made great strides in this space, but the reality is that many wheelchair users and transit drivers are unhappy with the current methods and opt to ignore securement. This is a tricky problem that still needs attention using a systems approach.

Our research also identified some important challenges to public transportation that are currently of concern to providers and planners caused by external factors like urban development patterns, demographics, health trends, and technology trends. Some of these may seem like they are outside the scope of this book but they are closely related in ways that influence outcomes in the accessible public transportation field.

The obesity epidemic requires transit agencies to accommodate much larger individuals and leads to increases in paratransit utilization. The impact will be most serious in the design and construction of boarding equipment like lifts and ramps. It also may impact design of paratransit vehicles. Due to the relationship between obesity and disabilities like those resulting from diabetes and the increase in childhood obesity, the prospects for the impact of this trend being reduced in the future do not look good.

The aging of the Boomer generation presents a challenge because older riders have different travel patterns, are often unfamiliar with public transportation, and have limitations in performance. Large numbers of this generation also live in low-density suburban and rural communities. Currently, public transit systems are designed around the journey to work. Service for those who do not work, therefore, is not up to par with expectations or need. To fill this gap, private "elder van" and targeted ride sharing services are emerging (ITNAmerica, 2015). These options create an interesting alternative to public transportation and also serve as a model for how public agencies, with their greater experience and resources, could expand their market. Private services are covered by the ADA, Title III although there are some differences compared to public entities that have to comply with Title II. One of the challenges associated with aging riders for public agencies is the costs of transportation to health care settings, a major destination of paratransit trips.

Serving rural areas is a subset of the broader Low-Density Challenge listed above. Rural areas have a triple challenge because they are much less likely to have any public transportation at all, have few resources to address the problem, and must provide service to areas with very low density. Lack of access to transportation is one of the most frequently cited problems for rural residents, particularly for people with disability living in rural areas (RTC Rural, 1999). Those who do have some form of public transportation are facing high costs combined with lowered government resources that threaten to eliminate service entirely. Innovative strategies are needed to lower costs and increase revenue from riders. One rural transportation agency in such a bind reported increasing ridership and saving money with a three-pronged strategy: (1) service routes were changed to reflect the shift in commercial development in the area; (2) comfortable bus stops at major shopping plazas were built in conjunction with retailers to make waiting comfortable; and, (3) former airport shuttles were purchased and equipped with lifts to replace large buses (DeRock, 2011). The lack of sidewalks is another important issue in rural areas since accessibility regulations require a paved loading platform at bus stops. A rural agency reported on the successful use of a low-cost paving product developed for military applications to build accessible bus stop platforms (DeRock, 2011).

Serving the suburbs has its own characteristic problems. Suburban communities often are served by metropolitan transit agencies but with only limited service, geared to commuters. Increasingly, urban low-income residents are commuting in the opposite direction to suburbanites who work in urban centers. Urban-based transit agencies have difficulty reaching the widely distributed work sites and other destinations in the suburbs.

Flex-route service with small vehicles providing last mile service to suburban stations along fixed-route lines may be a good solution to this problem. Partnerships between public agencies, private transportation providers like elder van companies, wheelchair van services, and taxi companies could result in the emergence of entirely new systems designed to serve the low-density suburban environment. A prototype of such a service is operating successfully in Australia. A not-for-profit company coordinates several types of transportation services, including volunteer drivers using their own cars, their own lift-equipped vans, to connections with fixed route public transportation. A triage approach is used to identify the specific needs of each individual and match them with the appropriate service. Further into the future, shared autonomous vehicles may become affordable, feasible, and economically viable for suburban service. Compared to dense urban areas and rural areas, suburban streets pose fewer challenges to robotic vehicles. Several states have passed laws that permit autonomous vehicles on public streets but policy issues remain. For example, some of these laws require a licensed driver behind the wheel of autonomous vehicles.

The impact of changes in energy policy on accessibility has not been studied. The recent increases in the price of oil are having a direct impact on transit operations but, over time, we can expect to see some impacts on accessibility as well. The most dramatic impact will probably come from high oil prices and increased adoption of transit-oriented development strategies. Both can lead to increased utilization of public transportation. This could affect accessibility in a positive way by increasing revenue to agencies and in a negative way by increasing crowding on transit vehicles. A less obvious impact is the shift to different fuel systems. Currently agencies are experimenting with "green buses" which run on natural gas. But, there is no difference in the design of these vehicles as far as passenger features are concerned. In the future, agencies will start exploring new technologies like electric power. This could lead to entirely different models of mass transportation, for example, the use of small one- and two-seat electric vehicles that are distributed around communities like bike sharing programs.

The digital divide is rapidly shrinking for the mobile space but is still a concern. However, smartphone ownership by adults with household incomes lower than $30,000 is increasing and this population is often smartphone dependent (Smith, 2015). A key factor to consider is that people in lower-income brackets may lack the resources to buy a home computer, but are able to purchase a smartphone, which subsequently becomes a main source of Internet access. While smartphones can now be obtained for free with two-year plans, the data plans for these devices can still be expensive. Unfortunately, many smartphone transit apps are data intensive. There is a need to find ways to manage data consumption so the technical and service advances provided by smartphones can be obtained with reduced data charges.

Accessible Public Transit from a Holistic Perspective: The Future of Public Transportation

In a country like the U.S., where most citizens, including policy makers, government officials, politicians, and researchers, use private automobile transportation exclusively, public transportation is often perceived as an obsolete and inferior mode of transportation. Their rare experiences using public transportation in the U.S. confirm their beliefs that comfort, security, and convenience can never be as good in public transportation as it is in a private automobile. Some public figures have even advocated disinvestment in public transportation as an ideological position (Center for American Progress, 2012;

Schweitzer, 2011). Even many individuals with disabilities try to avoid using public transportation, if they can. Thus, prejudiced attitudes are significant obstacles toward bringing more resources to bear on the issue of accessible public transportation.

Those who have experience with the very good service and environment of public transportation in other countries, especially Western Europe and Japan, know that public transportation can compete effectively with private automobiles if investment is sufficient to provide good service. Increasing knowledge about effective accessible public transportation practices in other countries, especially universal design approaches, should be an important goal for future education and dissemination activities. Not only do policy makers and the public need to be better informed about best practices, but transportation agencies could benefit greatly from such information. It would help them to adopt evidence-based practices and take advantage of practical knowledge throughout the world. A good example is the practice of locating the accessible entry in the front of buses in the U.S. As noted earlier, in Europe, the general practice is to locate the accessible entry in the rear or middle of the vehicle, between the axles. Although different cities in the U.S. adopted both approaches in the early history of accessible transportation, in recent years, the front location has become the norm. However, our research has discovered that this practice limits accessibility for wheeled mobility users.

Despite the negative attitudes toward public transportation in the U.S., congestion on highways, current economic conditions, and demographic changes are increasing ridership. The American Public Transportation Association (APTA) regularly releases national data on ridership. In 2010 it revealed that Americans took 10.2 billion trips on public transportation. Although this reflects a slight drop from 10.5 and 10.4 billion trips in 2008 and 2009, respectively, which maps to the recession, it still reflects a 31 percent overall increase in ridership since 1995 (APTA, 2012). AARP reports that 62 percent of adults aged 50 plus limited their daily driving when gas prices rose (AARP, 2005; Litman, 2006). Driving is not a viable option for many older persons. Nearly seven million persons 65 and older do not drive, due to either health or financial deterrents. Older adults with lower incomes are particularly effected by barriers in public transportation systems because they rely on public transportation more regularly (Houser, 2005). According to an analysis by APTA of 130 surveys conducted by transit systems, older people account for about 7.1 percent of trips on conventional transit services, or about 386 million trips per year (APTA, 2007). We can expect the number of people who will be dependent on public transportation to soar as the population ages. Interestingly, the younger generation is also cutting back on personal vehicle driving and many prefer to live in places conducive to walking, biking, and riding public transit (Davis, Dutzik, & Baxandall, 2012).

The trends in ridership suggest that the time is right to promote universal design. This means that accessibility features should be applied with an eye toward a varied population consisting of both people with and without disabilities. This also suggests rethinking all aspects of the transit environment to accommodate an aging ridership. There is a great deal of information on design for aging that could be mined and applied in the public transportation environment. In particular, our research shows that accessing and utilizing information on schedules and routes is a major problem for all user groups. We do not need new research to improve information and communication systems for the older population because general principles and specific strategies are already well known. But, we do need knowledge translation activities to make that information available to the product development community in a form that is useful.

Priorities for Accessible Public Transportation

We still believe that improving independent mobility in the community is currently a more important priority in the field than intercity travel. Public transit directly impacts community participation, including education, employment, recreation, spontaneous travel, and civic engagement. For similar reasons, we advocate a heavy focus on mainline public transit rather than paratransit. Having said this, many of the challenges and recommendations identified here are also relevant for intercity and paratransit applications, especially for vehicle design, the built environment, rider information, and service delivery.

Innovations and the application of research and best practices in this field must contribute to the outcomes of increased community participation including access to education facilities, health settings, employment, and civic engagement. Top-level priorities are:

Crosscutting Priority—Universal Design: *Universal design approaches and solutions are recommended for all priorities.* Our research highlights the value of cross-disability perspectives. Universal design is also important since people without disabilities will be in frequent and direct contact with accessibility solutions for vehicles, infrastructure, interaction experiences, and service delivery. Approaches that have value to riders of all capabilities are more likely to see implementation and be maintained over time because they directly affect service quality for all riders.

Crosscutting Priority—Stakeholder Involvement: *Effort to improve public transit accessibility needs clear plans for continuous involvement of riders, agencies, and industry.* Public transit consists of diverse and often disparate communities. A wide spectrum approach and partnerships with key stakeholder organizations are needed. Furthermore, the riders themselves are highly heterogeneous, thereby placing a premium on seeking input from a wide range of demographics in order to mitigate inadvertent bias.

Vehicle Design—Designs and technologies are needed to support rapid boarding and egress from public transit vehicles. Long bus stop dwell times associated with serving riders with mobility impairments are undesirable from both operations and rider acceptance perspectives. These can lead to pressure on drivers to violate agency policies and animosity toward riders with disabilities from both drivers and other riders. Advances have been made toward this goal, but more work is still needed both in advancing research and the application of recently generated knowledge. Simple design changes of one component, like the pedestal of fare machines, may contribute to accessibility just as much as design strategies that are more difficult to implement, like widening the space between wheel wells.

Rider Information—Technology that provides real-time information for both regular and spontaneous use of public transit. Riders with disabilities need to know if a vehicle is coming, when it will arrive, whether they will be able to board it, and whether their destination is accessible. Although much of this can be provided through existing technology, it results in very high costs and requires custom software integration. Solutions that lower these costs and streamline delivery of this information and ensure accessibility of information are needed.

Service Interaction—Technology that advances riders' ability to influence and enhance public transit service delivery. Traditional models for rider-agency interaction are slow, constrained, bureaucratic, and often inherently antagonistic. Riders are sources of knowledge and expertise beyond simple documentation of ADA violations and other problems. New methods for engaging riders through co-design, co-production, and other modern service models are needed. These will increase collaboration and dialog between riders and agencies, thereby streamlining progress toward better service quality.

Infrastructure and the "Last Mile"—Solutions are needed to help riders with disabilities access stops and stations. Even if transit vehicles allow rapid boarding and good service policies are followed, a rider with disabilities may be blocked from accessing public transit by infrastructure barriers at the start, transfers, and the end of their trip. Information technology strategies are needed to help riders be aware of and respond to physical barriers before boarding vehicles and after disembarking. There is a need for innovation and deployment of methods for evaluating accessibility of rights-of-way surrounding stations and, where there are no pedestrian access routes, approaches to on-demand and other personalized services to get people to the stop.

In closing, there has been great progress toward efficient and successful delivery of accessible transportation, especially since the enactment of the ADA. Research has also identified the potential benefits of consumer-oriented strategies. However, there are increasing economic and demographic pressures on these advances and future improvements. It is critical that advances in accessible public transportation be cognizant of the larger challenges experienced by service providers and that accessibility issues extend beyond the scope of current standards. Whenever possible, practitioners and researchers should seek out universal design approaches that leverage the larger market forces and society demographics. Universal design standards for transportation systems will raise awareness and provide tools for practice.

References

AARP. (2005). *High gas prices cause 50+ to modify lifestyle.* AARP Bulletin Survey/News Release. Washington, DC: AARP Public Policy Institute.

AccessMyNYC. (2012). Retrieved from www-03.ibm.com/able/news/Accessmynyc.html (Accessed November 16, 2015).

Agrawal, A. M., Schlossberg, M., & Irvin, K. (2008). How far, by which route, and why? A spatial analysis of pedestrian preference. *Journal of Urban Design, 13*(1), 81–98.

Albertson, P., & Falkmer, T. (2005). Is there a pattern in European bus and coach incidents? A literature analysis with special focus on injury causation and injury mechanisms. *Accident Analysis and Prevention, 37*, 225–233.

Alsnih, R., & Hensher, D. (2003). The mobility and accessibility expectations of seniors in an aging population. *Transportation Research A, 37*, 903–916. doi:10.1016/S0965-8564(03)00073-9.

Amini, S., Lindqvist, J., Hong, J. I., Lin, J., Toch, E., & Sadeh, N. M. (2011). Caché: Caching location-enhanced content to improve user privacy. In *Proceedings of the International Conference on Mobile Systems, Applications, and Services (MobiSys)*, pp. 197–210.

ANSI/RESNA. (2000). *ANSI/RESNA WC-19: Wheelchairs used as seats in motor vehicles (Wheelchair Standard).* American National Standards Institute/Rehabilitation Engineering and Assistive Technology Society of North America. Washington, DC: RESNA.

APTA. (2007a, May). *A profile of public transportation passenger demographics and travel characteristics reported in on-board surveys.* Washington, DC: American Public Transportation Association.

APTA. (2009). *Standard bus procurement guidelines RFP.* Draft 12/11/09. Washington, DC: American Public Transportation Association.

APTA. (2012). *Public transit ridership up in the second quarter marking the sixth consecutive quarter of ridership growth.* Retrieved from www.apta.com/mediacenter/pressreleases/2012/Pages/120918_2ndQuarterRidership.aspx.

Bailey, L. (2004). *Aging Americans: Stranded without options.* Surface Transportation Policy Project. Retrieved from www.apta.com/resources/reportsandpublications/Documents/aging_strande d.pdf.

Barbeau, S., Georggi, N., & Winters, P. (2010). Integration of GPS-enabled mobile phones and AVL: Personalized real-time transit navigation information on your phone. In *Proceedings from Transportation Research Board 2010 Annual Meeting.*

Barbulescu, L., Rubinstein, Z., Smith, S. F., & Zimmerman, T. L. (2010, May). Distributed co-ordination of mobile agent teams: The advantage of planning ahead. In *Proceedings of the 9th International Conference on Autonomous Agents and Multi-Agent Systems*, Toronto, CA.

Bareria, P., D'Souza, C., Lenker, J., Paquet, V., & Steinfeld, E. (2012). *Performance of visually impaired users during simulated boarding and alighting on low-floor buses.* Paper presented at the Human Factors and Ergonomics Society 56th Annual Meeting, October 2012, Boston, MA.

Baum-Snow, N. (2007). Did highways cause suburbanization? *Quarterly Journal of Economics, 122*(2), 775–805.

Bednar, M. (1977). *Barrier free environments.* Stroudsburg, PA: Dowden, Hutchinson, and Ross.

Bentzen, B. L., Nolin, T. L., Easton, R. D., Desmarais, L., & Mitchell, P. A. (1994). *Detectable warning surfaces: Detectability by individuals with visual impairments, and safety and negotiability for individuals with physical impairments. Final report DOT-VNTSC-FTA-94-4 and FTA-MA-06-0201-94-2.* Washington, DC: U.S. Department of Transportation, Federal Transit Administration, Volpe National Transportation Systems Center, and Project ACTION, National Easter Seal Society.

Berbeglia, G., Pesant, G., & Rousseau, L.-M. (2011). Checking the feasibility of dial-a-ride instances using constraint programming. *Transportation Science, 45,* 399–412.

Biagioni, J., Agresta, A., Gerlich, T., & Eriksson, J. (2009). TransitGenie: A context-aware, real-time transit navigator. In *Proceedings from SenSys '09: 7th ACM Conference on Embedded Networked Sensor Systems,* Berkeley, CA and New York.

Black, A., Burger, S., Conkie, A., Hastie, H., Keizer, S., Lemon, O., ..., Eskenazi, M. (2010). Spoken dialog challenge: Comparison of live and control test results. *Annual Meeting of the Special Interest Group on Discourse and Dialogue (SIGDial),* pp. 2–7.

Blennemann, F. (1991). The low floor bus concept: Advantages for the elderly and handicapped. *Planning and Transport Research and Computation, P349,* 29–42.

Boyne, G. A. (July 2003). Sources of public service improvement: A critical review and research agenda. *Journal of Public Administration Research and Theory, 13*(3), 367–394.

Brabham, D., Sanchez, T., & Bartholomew, K. (2010). Crowd-sourcing public participation in transit planning: Preliminary results from next stop design case. *Transportation Research Board 2010 Annual Meeting.*

Bradwell, P., & Marr, S. (2008). *Making the most of collaboration: An international survey of public service co-design, DEMOS report 23.* DEMOS, in association with PriceWaterhouseCoopers (PWC) Public Sector Research Centre, London.

Brakewood, C., Macfarlane, G. S., & Watkins, K. (2015). The impact of real-time information on bus ridership in New York City. *Transportation Research Part C: Emerging Technologies, 53,* 59–75.

Bridj. (2015). *How it works.* Retrieved from www.bridj.com/#how.

Brooks, D. (2015). Driverless cars in the near future for Fort Bragg. *Fayetteville Observer.* Retrieved from www.fayobserver.com/news/local/driverless-cars-in-the-near-future-for-fort-bragg/article_9ab20600-c91a-5360-b1a4-be8bcc90dce1.html.

Brown, S. R. (September 16, 2014). Judge approves city's push for wheelchair-accessible cabs and 30-cent surcharge. *New York Daily News.* Retrieved from www.nydailynews.com/new-york/judge-approves-push-wheelchair-accessible-cabs-30-cent-surcharge-article-1.1941591.

Buning, M. E., Getchell, C. A., Bertocci, G. E., & Fitzgerald, S. G. (2007). Riding a bus while seated in a wheelchair: A pilot study of attitudes and behavior regarding safety practices. *Assistive Technology, 19*(4), 166–179.

Bureau of Transportation Statistics, U.S. Department of Transportation. (2003). *Transportation difficulties keep over half a million disabled at home, April 2003,* BTS Issue Brief, No. 3. Retrieved from www.bts.gov/publications/issue_briefs/number_03.

Burke, J. A., Estrin, D., Hansen, M., Parker, A., Ramanathan, N., Reddy, S., & Srivastava, M. B. (2006). Participatory sensing. In *Workshop on World-Sensor-Web.*

Burns, L. D. (2013). Sustainable mobility: A vision of our transport future. *Nature, 497,* 181–182.

Burns, L. D., Jordan, W. C., & Scarborough, B. A. (2013). *Transforming personal mobility.* Earth Institute, Columbia University. Retrieved from http://sustainablemobility.ei.columbia.edu/files/2012/12/Transforming-Personal-Mobility-Jan-27-20132.pdf.

Cannon, D. (2011). Panel discussion on "Building a Global Community of Practice Around Accessible Transportation: How to Create a Foundation for Evidence-Based Practice", at the Transportation Research Board 90th Annual Meeting, Washington, DC, January 23, 2011.

Carma. (2015). From www.gocarma.com/.

Casey, C. (2003). Real-time information: Now arriving. *Metro Magazine,* April Issue.

Center for American Progress. (2012). *New Ryan budget disinvests in America.* Retrieved from www.americanprogress.org/issues/budget/news/2012/03/20/11340/new-ryan-budget-disinvests-in-america/.

Cervero, R. (2000). *Informal transport in the developing world.* Nairobi: United Nations Centre for Human Settlements (Habitat).

Chango, M. (2007). Challenges to e-government in Africa south of Sahara. In *Proceedings of the 1st International Conference on Theory and Practice of Electronic Governance.*

Cho, J., & Chun, S. A. (2011). Towards transparent policy decision making process. In *Proceedings of the 12th Annual International Digital Government Research Conference: Digital Government Innovation in Challenging Times.*

Christophersen, J. (Ed.). (2002). *Universal design: 17 ways of thinking and teaching.* Oslo: Husbanken.

CityMobil2. (2015). CityMobil2. Retrieved from www.citymobil2.eu/en/.

City of New York. (2016a). About TLC. Retrieved from www.nyc.gov/html/tlc/html/about/about.shtml.

City of New York. (2016b). Taxi of Tomorrow. Retrieved from www.nyc.gov/html/media/totweb/taxioftomorrow_home.html.

Collia, D., Sharp, J., & Giesbrecht, L. (2003). The 2001 National Household Travel Survey: A look into the travel patterns of older Americans. *Journal of Safety Research, 34,* 461–470. doi:10.1016/j.jsr.2003.10.001.

Connell, B., Jones, M., Mace, R., Meuller, J., Mullick, A., Ostroff, E. ... Vanderheiden, G. (1997). *The principles of universal design.* Retrieved from www.design.ncsu.edu/cud/univ_design/principles/udprinciples.htm.

Constandache, I., Gaonkar, S., Sayler, M., Choudhury, R. R., & Cox, L. P. (2009). EnLoc: Energy-efficient localization for mobile phones. *IEEE INFOCOM – INFOCOM,* pp. 2716–2720.

Cooper, R. A., Cooper, R., & Boninger, M. L. (2008). Trends and issues in wheelchair technologies. *Assistive Technology, 20*(2), 61–72.

Cordeau, J.-F. (2006). A branch-and-cut algorithm for the dial-a-ride problem. *Operations Research, 54*(3), 573–586.

Cordeau, J.-F., & Laporte, G. (2007). The dial-a-ride problem: Models and algorithms. *Annals of Operations Research, 153*(1), 29–46.

Cowan, S. (2014). *Uber and Lyft are crushing the taxi companies the disabled rely on.* TakePart. Retrieved from www.takepart.com/article/2014/09/17/uber-and-lyft-crush-taxicompanies-disabled-rely

Crandall, W., Brabyn, J., Bentzen, B. L., & Myers, L. (1999). Remote infrared signage evaluation for transit stations and intersections. *Journal of Rehabilitation Research and Development, 36*(4), 341–355.

Cross, D. J. (2011). *Mobility device securement: Standards and wheelchair marking & tether strap programs.* Oakland, CA: Douglas J. Cross Transportation Consulting. Retrieved October 17, 2012, from www.douglasjcross.com/Cross_WC_StdsMarkingTether_R6_APTA_Oct2011.pdf.

D'Souza, C., Paquet, V., Lenker, J. A., & Steinfeld, E. (2017). Effects of transit bus interior configuration on performance of wheeled mobility users during simulated boarding and disembarking. *Applied Ergonomics, 62,* 94–106.

D'Souza, C., Paquet, V., Lenker, J., Steinfeld, E., & Bareria, P. (2012). Low-floor bus design preferences of walking aid users during simulated boarding and alighting. *Work: A Journal of Prevention, Assessment and Rehabilitation, 41*(Supplement 1), 4951–4956.

Daamen, W., de Boer, E., & de Kloe, R. (2008). Assessing the gap between public transport vehicles and platforms as a barrier for the disabled: Use of laboratory experiments. *Transportation Research Record: Journal of the Transportation Research Board, 2072,* 131–138.

Dabson, B., Johnson, T. G., & Fluharty, C. W. (2011). *Rethinking federal investments in rural transportation: Rural considerations regarding reauthorization of the Surface Transportation Act.* Columbia, MO: Rural Policy Research Institute.

Danford, G. S., & Maurer, J. (2005). Empirical tests of the claimed benefits of universal design. In *Proceedings of the Thirty-Sixth Annual International Conference of the Environmental Design Research Association.* Edmond, OK: Environmental Design Research Association, pp. 123–128.

Davies, J., Janowski, T., Ojo, A., & Shukla, A. (2007). Technological foundations of electronic governance. In *Proceedings of the International Conference on Theory and Practice in Electronic Governance.*

Davis, B., Dutzik, T., & Baxandall, P. (2012). *Transportation and the new generation: Why young people are driving less and what it means for transportation policy.* Washington, DC: Frontier

Group and U.S. PIRG Education Fund. Retrieved from www.uspirgedfund.org/reports/usp/transportation-and-new-generation (Accessed September 25, 2012).

Davison, L., Enoch, M., Ryley, T., Quddus, M., & Wang, C. (2014). A survey of demand responsive transport in Great Britain. *Transport Policy, 31*, 47–54.

DeRock, R. A. (2011). *Cost-effective accessibility improvements in small urban and rural environment.* Paper presented at the Transportation Research Board, January 23–27, 2011, Washington, DC.

DialRC. (2012). Retrieved from http://dialrc.org (Accessed August 16, 2012).

Disability Rights Education & Defense Fund. (2010). No-Shows in ADA Paratransit. *A series of topic guides providing technical assistance for transit agencies, riders and advocates on the Americans with Disabilities Act (ADA) and transportation.* Retrieved from http://dredf.org/ADAtg/noshow.shtml.

Dwyer, M. C., & Beavers, K. A. (2011). *Economic vitality: How the arts and culture sector catalyzes economic vitality.* Arts and Culture Briefing Papers 05. Chicago, IL: American Planning Association.

Dziekan, K., & Kottenhoff, K. (2007). Dynamic at-stop real-time information displays for public transport: Effects on customers. *Transportation Research Part A: Policy and Practice, 41*(6), 489–501.

Easter Seals Project Action. (2006). *Toolkit for the assessment of bus stop accessibility and safety.* Washington, DC: Easter Seals Project Action.

Easter Seals Project Action. (2009). *Universal design & accessible transit systems: Facts to consider when updating or expanding your transit system.* Retrieved from www.easterseals.com/site/Ecommerce Download/Universal_Design_FactSheet-5821.pdf?dnl=90752-5821-761N6ivu74JPUrFe.

Ecolane. (2015). Ecolane. Retrieved from www.ecolane.com.

Edvardsson, B., Gustafsson, A., & Roos, I. (2005). Service portraits in service research: A critical review. *International Journal of Service Industry Management, 16*(1), 107–121.

Enoch, M., Potter, S., Parkhurst, G., & Smith, M. (2004). *Intermode: Innovations in demand responsive transport.* Loughborough: Loughborough University, Department for Transport and Greater Manchester Passenger Transport Executive.

Eppli, M. J., & Tu, C. C. (1999). *Valuing the new urbanism: The impact of the new urbanism on prices of single-family homes.* Washington, DC: ULI-the Urban Land Institute.

Equal Rights Center. (2007). Allegations of discrimination lodged against two popular car-sharing companies—Flexcar and Zipcar. Retrieved from www.equalrightscenter.org/.

Fagnant, D. J., & Kockelman, K. M. (2015). Dynamic ride-sharing and optimal fleet sizing for a system of shared autonomous vehicles. In *Proceedings of the Annual Meeting of the Transportation Research Board.* Retrieved from www.ce.utexas.edu/prof/kockelman/public_html/TRB15SAVswithDRSinAustin.pdf.

Faulring, A., Myers, B., Mohnkern, K., Schmerl, B., Steinfeld, A., Zimmerman, J., …, Siewiorek, D. (2010). Agent-assisted task management that reduces email overload. In *Proceedings of the ACM International Conference on Intelligent User Interfaces (IUI).*

Fernández, R., Zegers, P., Weber, G., & Tyler, N. (2010). Influence of platform height, door width, and fare collection on bus dwell time. *Transportation Research Record: Journal of the Transportation Research Board, 2143*(1), 59–66.

Ferris, B., Watkins, K., & Borning, A. (2010a). OneBusAway: Location-aware tools for improving public transit usability. *IEEE Pervasive Computing, 9*(1), 13–19.

Ferris, B., Watkins, K., & Borning, A. (2010b). OneBusAway: Results from providing real-time arrival information for public transit. In *Proceedings from the 28th International Conference on Human Factors in Computing Systems (CHI),* Atlanta, GA; New York, pp. 1807–1816.

Fitzgerald, S. G., Songer, T., Rotko, K., & Karg, P. (2007). Motor vehicle transportation use and related adverse events among persons who use wheelchairs. *Assistive Technology, 19*(4), 180–187.

Flegenheimer, M. (2012, June 28). Court says taxi system complies with disabilities law. *The New York Times, Region.* Retrieved from www.nytimes.com/2012/06/29/nyregion/new-york-citys-taxi-system-complies-with-disabilities-law-panel-rules.html.

Florida Alliance for Assistive Services and Technology. (2012). *General self-help resource guide: Access to rental cars, other vehicles, and lodging accommodations to assist individuals with disabilities under the ADA and ADA Amendments Act of 2008.*

Florida, R. (2002). *The rise of the creative class.* New York: Basic Books.

Foot, K. (2007). Keynote speech at the ITE 2007 Annual Meeting and Exhibit. March 25–28, 2007, San Diego, CA.

Forlizzi, J., & Zimmerman, J. (2013). *Promoting service design as a core practice in HCI.* In Submission to the ACM Conference on Human Factors in Computing Systems, ACM Press.

Freed, M., Carbonell, J., Gordon, G., Myers, B., Siewiorek, D., Smith, S., ..., Tomasic, A. (2008). RADAR: A personal assistant that learns to reduce email overload. In *Proceedings of the AAAI Conference on Artificial Intelligence*, pp. 1287–1293.

Frieden, L. (2005). *The current state of transportation for people with disabilities in the United States.* Washington, DC: National Council on Disability.

Frost, K., & Bertocci, G. (2007). *Wheelchair rider incidents on public transit buses: A 4-year retrospective review of metropolitan transit agency records.* Paper presented at the 28th Annual RESNA Conference, Phoenix, AZ.

Frost, K. L., & Bertocci, G. (2010). Retrospective review of adverse incidents involving passengers seated in wheeled mobility devices while traveling in large accessible transit vehicles. *Medical Engineering & Physics, 32*, 230–236.

Frost, K. L., Bertocci, G., & Salipur, Z. (2013). Wheelchair securement and occupant restraint system (WTORS) practices in public transit buses. *Assistive Technology, 25*(1), 16–23.

Frost, K. L., Bertocci, G. E., & Sison, S. (2010). Ingress/egress incidents involving wheelchair users in a fixed-route public transit environment. *Journal of Public Transportation, 13*(4), 41–62.

Frost, K. L., Bertocci, G., & Smalley, C. (2015). Ramp-related incidents involving wheeled mobility device users during transit bus boarding/alighting. *Archives of Physical Medicine & Rehabilitation, 96*, 928–933.

Frost, K. L., van Roosmalen, L., Bertocci, G. E., & Cross, D. (2012). Wheeled mobility device transportation safety in fixed route and demand-responsive public transit vehicles within the United States. *Assistive Technology, 24*(2), 87–101.

Gardiner, S., Tomasic, A., Zimmerman, J., Aziz, R., & Rivard, K. (2011). Mixer: Mixed-initiative data retrieval and integration by example. *Human-Computer Interaction-INTERACT 2011–13th IFIP TC 13 International Conference*, pp. 426–443.

Gates, R., & Kalanick, T. (2014). How Uber is helping veterans. *Politico Magazine.* Retrieved from www.politico.com/magazine/story/2014/09/robert-gates-uber-veterans-111039.html#.VZvk9m D9OiX.

Gaver, W. W., Dunne, T., & Pacenti, E. (1999). Cultural probes. *Interactions, 6*(1), 21–29.

Geo-Access Challenge Team. (2011). *Data-enabled travel: How geo-data can support inclusive transportation, tourism, and navigation through communities.*

Gildea, G., & Sheikh, M. (1996). Applications of technology in providing transit information. *Transportation Research Record, 1521*, 71–76.

Goldman, J. M., & Murray, G. (2011). *TCRP Synthesis 88: Strollers, carts, and other large items on buses and trains.* Washington, DC: Transportation Research Board.

Goldstein, S. M., Johnston, R., Duffy, J. A. & Rao, J. (2002). The service concept: The missing link in service design research? *Journal of Operations Management, 20*(2), 121–134.

Golledge, R., Costanzo, C. M., & Marston, J. (1996). Public transit use by non-driving disabled persons: The case of the blind and vision impaired. Retrieved from http://repositories.cdlib.org/its/path/papers/UCB-ITS-PWP-96-1.

Golledge, R. G., Marston, J. R., & Costanzo, C. M. (1998). *Assistive devices and services for the disabled: Auditory signage and the accessible city for blind or vision impaired travelers. Working Paper UCB-ITS-PWP-98-18.* Berkeley: PATH Program, Institute for Transportation Studies, University of California, Berkeley.

Google. (2012). *Self-driving car test: Steve Mahan.* YouTube. Retrieved from www.youtube.com/watch?v=cdgQpa1pUUE.

Google. (2014). *Just press go: Designing a self-driving vehicle.* Retrieved from https://googleblog.blogspot.com/2014/05/just-press-go-designing-self-driving.html.

Harper, C., Hendrickson, C., Mangones, S., & Samaras, C. (2016). Estimating potential increases in travel with autonomous vehicles for the non-driving, elderly and people with travel restrictive medical conditions. *Transportation Research Part C: Emerging Technologies, Volume 72*: 1–9. http://dx.doi.org/10.1016/j.trc.2016.09.003

Hauger, J., Rigby, J., Safewright, M., & McAuley, W. (1996). Detectable warning surfaces at curb ramps. *Journal of Visual Impairments and Blindness, 90*, 512–525.

Hendershot, G. (2003). *Community participation and life satisfaction.* Retrieved from www.nod. org/index.cfm?fuseaction=Feature.showFeature&FeatureID=112 9 (Accessed March 12, 2008).

Hess, D. B. (2009). Access to public transit and its influence on ridership for older adults in two U.S. cities. *Journal of Transport and Land Use, 2*(1), 3–27.

Hobson, D. A., & van Roosmalen, L. (2007). Towards the next generation of wheelchair securement—development of a demonstration UDIG-compatible wheelchair docking device. *Assistive Technology, 19*(4), 210–222.

Houser, A. (2005). *Community mobility options: The older person's interest.* Washington, DC: AARP Public Policy Institute.

Howe, J. (2006). *The rise of crowdsourcing.* Retrieved from www.wired.com/wired/archive/14.06/crowds_pr.html (Accessed September 12, 2012).

Indian Trails. (2012). *Indian Trails and MDOT launch first U.S. bus fleet with 'hearing loop' technology.* Retrieved from www.prnewswire.com/news-releases/indian-trails-and-mdot-launch-first-us-bus-fleet-with-hearing-loop-technology-159608075.html (Accessed August 16, 2012).

International Code Council (ICC). (2009). *2009 international building code.* Country Club Hills, IL: International Code Council.

International Disability Rights Monitor. (2004). *IDRM: Regional report of the Americas.* Chicago, IL: Center for International Rehabilitation. Retrieved from bbi.syr.edu/publications/blanck_docs/2003-2004/IDRM_Americas_2004.pdf.

International Disability Rights Monitor. (2005). *IDRM: Regional report of Asia 2005.* Chicago, IL: Center for International Rehabilitation. Retrieved from www.idrmnet.org/content.cfm?id=5E5A75&m=3.

International Disability Rights Monitor. (2007a). *IDRM: Regional report of the Americas 2004.* Chicago, IL: Center for International Rehabilitation. Retrieved from www.idrmnet.org/content.cfm?id=5E5A75&m=3.

International Disability Rights Monitor. (2007b). *IDRM: Regional report of Europe 2007.* Chicago, IL: Center for International Rehabilitation. Retrieved from www.idrmnet.org/content.cfm?id=5E5A75&m=3.

International Road Transport Union (IRTU). (2007). *Improving access to taxis.* European Conference of Ministers of Transport. Retrieved from www.internationaltransportforum.org/pub/pdf/07TaxisE.pdf.

ITNAmerica. (2015). *Independent transportation network of America.* Retrieved from www.itnamerica.org.

Iwarsson, S., Jensen, G. & Stahl, A. (2000). Travel chain enabler: Development of a pilot instrument for assessment of urban public bus transport accessibility. *Technology and Disability, 12*, 3–12.

Jaffe, E. (2014). *Lyft is hiring a lot of deaf drivers.* Citylab. Retrieved from www.citylab.com/work/2014/09/lyft-is-quietly-hiring-a-lot-of-deaf-drivers/380672/.

Jain, S., & Van Hentenryck, P. (2011). Large neighborhood search for dial-a-ride problems. In *Proceedings of the 17th International Conference on Principles and Practice of Constraint Programming.* Berlin/Heidelberg: Springer, pp. 400–413.

Johnson, K. (2006). Demographic trends in rural and small town America. *Reports on Rural America, 1*(1), 1–35. Durham, NH: University of New Hampshire, Carsey Institute.

Johnson, N., Davis, T., & Bosanquet, N. (2000). The epidemic of Alzheimer's disease: How can we manage the costs? *PharmacoEconomics, 18*(3), 215–223.

Kanbayashi, A. (1999). Accessibility for the disabled. *Japan Railway and Transport Review, 20*, 22–24.

Karg, P., Buning, M. E., Bertocci, G., Furhman, S., Hobson, D., & Manary, M. A. (2009). State of the science workshop on wheelchair transportation safety. *Assistive Technology, 21*(3), 115–160.

Kaschesky, M., & Riedl, R. (2009). Top-level decisions through public deliberation on the internet. In *Proceedings of the 10th Annual International Conference on Digital Government Research: Social Networks: Making Connections between Citizens, Data and Government.*

Kawauchi, Y. (1999). Railway stations and the right to equality. *Japan Railway and Transport Review, 20.*

Kehret, G., Miele, J., & Landau, S. (2011). *Development of smartpen-based audio/tactile transit station maps for travel planning and wayfinding.* CSUN Technology and Persons with Disabilities Conference, San Diego, CA, March 2011.

Kim, D. H., Kim, Y., Estrin, D., & Srivastava, M. B. (2010). SensLoc: Sensing everyday places and paths using less energy. In *Proceedings of the ACM Conference on Embedded Networked Sensor Systems (SenSys).*

King, S. F., & Brown, P. (2007). Fix my street or else: Using the internet to voice local public service concerns. In *Proceedings of Theory and Practice of Electronic Governance.* ACM Press, pp. 72–80.

Kirchner, C. E., Gerber, E. G., & Smith, B. C. (2008). Designed to deter. Community barriers to physical activity for people with visual or motor impairments. [Research Support, Non-U.S. Gov't]. *American Journal of Preventive Medicine, 34*(4), 349–352.

Koffman, D., Raphael, D., & Weiner, R. (2004). *The impact of federal programs on transportation for older adults (2004–17).* Washington, DC: AARP Public Policy Institute.

Koontz, A. M., Roche, B. M., Collinger, J. L., Cooper, R. A., & Boninger, M. L. (2009). Manual wheelchair propulsion patterns on natural surfaces. *Archives of Physical Medicine and Rehabilitation, 90*(11), 1916–1923.

Kourmpanis, V., & Peristeras, V. (2010). E-consultation: Processing and summarizing contributions with argument maps. In *Proceedings of the 4th International Conference on Theory and Practice of Electronic Governance, 2010.*

Kuneida, M., & Roberts, P. (2006). *Inclusive access and mobility in developing countries.* Washington, DC: World Bank. Retrieved from http://siteresources.worldbank.org/INTTSR/Resources/07-0297.pdf.

Kwong, J. (September 16, 2014). Report says SF taxis suffering greatly. *The Examiner.* Retrieved from http://archives.sfexaminer.com/sanfrancisco/report-says-sf-taxis-suffering-greatly/Content?oid=2899618.

Lane, J. P., & Flagg, J. L. (2010). Translating three states of knowledge—Discovery, invention, and innovation. *Implementation Science, 5*(9). doi: 10.1186/1748-5908-5-9.

LaPlante, M. P., & Kaye, H. S. (2010). Demographics and trends in wheeled mobility equipment use and accessibility in the community. *Assistive Technology, 22,* 3–17.

Lazlo, L. (2016). Uber flirts with transit agencies across the U.S. for a share of paratransit services. *The Washington Post.* Retrieved from www.washingtonpost.com/local/trafficand commuting/uber-flirts-with-transit-agencies-across-the-us-for-a-share-of-paratransit-services/2016/03/05/5eb8b118-d751-11e5-9823-02b905009f99_story.html.

Lenhart, A. (2015). *Teens, social media & technology overview 2015.* Pew Research Center. Retrieved from www.pewinternet.org/2015/04/09/teens-social-media-technology-2015/.

Lenker, J. A., Damle, U., D'Souza, C., Paquet, V., Mashtare, T., & Steinfeld, E. (2016). A usability evaluation of access ramps in public transit buses. *Journal of Public Transportation, 19*(2), 109–127.

Let's Go!. Retrieved from www.speech.cs.cmu.edu/letsgo (Accessed August 16, 2012).

Li, C., & Willis, K. (2006). Modeling context aware interaction for wayfinding using mobile devices. In *Proceedings from Mobile HCI.*

Liljas, P. (2014). Google's new car doesn't have a steering wheel. *Time.com.* Retrieved from http://time.com/121680/google-self-driving-car-autonomous-vehicle-advanced-prototype.

Lim, A., & Zhang, X. (2007). A two-stage heuristic with ejection pools and generalized ejection chains for the vehicle routing problem with time windows. *INFORMS Journal on Computing, 19*(3), 443–457.

Lin, K., Kansal, A., Lymberopoulos, D., & Zhao, F. (2010). Energy-accuracy trade-off for continuous mobile device location. In *Proceedings of the ACM International Conference on Mobile Systems, Applications, and Services (MobiSys).*

LINC Design LLC. (2012). *The BusBuddy-A solution for fixed route transit.* Last updated July 13, 2012. Retrieved September 23, 2012, from www.linc-design.com/LINC/Projects/Entries/2012/7/13_The_BusBuddy.html.

Litman, T. (2006). *Responses to "A desire named streetcar"*. Vancouver, BC: Victoria Transport Policy Institute.

Lobjois, R., & Cavallo, V. (2009). The effects of aging on street-crossing behavior: From estimation to actual crossing. *Accident Analysis and Prevention, 41*(2), 259–267.

London Taxi Company. (2016). Retrieved from http://london-taxis.co.uk/.

Loprest, P., & Maag, E. (2001). *Barriers to and supports for work among adults with disabilities: Results from the NHIS-D*. Washington, DC: The Urban Institute.

Lusher, R. H., & Mace, R. (1989). Design for physical and mental disabilities. In J. A Wilkes & R. T. Packard (Eds.), *Encyclopedia of architecture: Design engineering and construction*. New York: John Wiley and Sons, pp. 748–763.

Lyft. (2017). *Policies & other info*. Retrieved from https://help.lyft.com/hc/en-us/categories/201234918-Policies-Other-Info.

Mace, R. (1985). *Universal design: Barrier free environments for everyone*. Los Angeles, CA: Designers West.

Machado, S., Jose, R., & Moreira, A. (2012). *Social interactions around public transportation*. Information Systems and Technologies (CISTI), 2012 7th Iberian Conference, pp. 1–6, 20–23 June 2012.

Masood, M., & Nicholas, L. (2003). An empirical study of textual and graphical travel itinerary visualization using mobile phones. In *Proceedings from Australasian User Interface Conference*.

Matheus, R., & Ribeiro, M. M. (2009a). Public online consultation of federal ministries and federal regulatory agencies in Brazil. In *Proceedings of the 3rd International Conference on Theory and Practice of Electronic Governance*.

Matheus, R., & Ribeiro, M. M. (2009b). Models for citizen engagement in Latin American. In *Proceedings of the 3rd International Conference on Theory and Practice of Electronic Governance*.

Maynes Associates. (2009). *Safety & security at suburban bus stops*. Pittsburgh, PA: Airport Corridor Transportation Association.

Miele, J., & Landau, S. (2010). *Audio-tactile interactive computing with the livescribe pulse pen*. CSUN Technology and Persons with Disabilities Conference, San Diego, CA, March 2010.

Miller, M. (2012). *Remote infrared audible signage pilot program*. FTA Report No. 0012. Federal Transit Administration, U.S. Department of Transportation. Retrieved from www.fta.dot.gov/documents/FTA_Report_No._0012.pdf.

Mollenkopf, H., Bass, S., Kaspar, R., Oswald, R., & Wahl, H. (2006). Outdoor mobility in late life: Persons, environments and society. *The Many Faces of Health, Competence and Well-Being in Old Age, I*, 33–45.

Morton, T., & Yousuf, M. (2011). *Technological innovations in transportation for people with disabilities workshop summary report*. FHWA-HRT-11-041. Washington, DC: Office of Operations Research and Development, Federal Highway Administration, U.S. Department of Transportation.

Musil, S. (2012). Samsung smartphones vulnerable to remote data wipe. *CNET*. Retrieved from http://news.cnet.com/8301-1009_3-57520327-83/samsung-smartphones-vulnerable-to-remote-data-wipe/ (Accessed September 25, 2012).

MV1. www.mv-1.us/ (Accessed October 28, 2016).

Nabors, D. (2009). Federal Highway Administration Pedestrian Safety Guidance for Transit. Pedestrian and Bicycle Information Center Livable Communities Webinar Series.

Nagata, Y., & Brysy, O. (2009). A powerful route minimization heuristic for the vehicle routing problem with time windows. *Operations Research Letters, 37*(5), 333–338.

National Capital Region Transportation Planning Board. (2015). Commuter connections. Retrieved from www.commuterconnections.org/.

National Council on Disability. (2005). *The current state of transportation for people with disabilities in the United States*. Washington, DC: National Council on Disability.

National Highway Traffic Safety Administration. (1997). *Wheelchair-users' injuries and deaths associated with motor vehicle related incidents*. Washington, DC: National Highway Traffic Safety Administration.

National Organization on Disability. (2000). *N.O.D./Harris survey of community participation, 2000*. Retrieved from www.nod.org/content.cfm?id=798 (Accessed March 4, 2008).

National Organization on Disability. (2004). *N.O.D./Harris survey of Americans with disabilities.* Retrieved from http://nod.org/what_we_do/research/surveys/harris/.

NavCog. (2015). NavCog. Retrieved from www.cs.cmu.edu/~NavCog/ (Accessed October 15, 2015).

NavPal. (2012). *A smart phone-based navigation aid for blind and deafblind users.* Retrieved from www.cs.cmu.edu/~navpal (Accessed November 16, 2015).

NCAM. (2012). *Accessibility review of transit-related applications.* Boston, MA: National Center for Accessible Media at WGBH.

NCSL. (2014, October/November). *Getting there: NCSL's mobility newsletter,* Vol. 1: No. 4. Retrieved from www.ncsl.org/Portals/1/Documents/transportation/NCSLGettingThere-Oct-Nov2014. pdf.

Nelson\Nygaard Consulting Associates. (2008). *Status report on the use of wheelchairs and other mobility devices on public and private transportation.* Washington, DC: Easter Seals Project ACTION. TRB workshop.

Newman, O. (1996). *Creating defensible space.* Washington, DC: U.S. Department of Housing and Urban Development (Contract No. DU100C000005967).

Norman, D. (2007). *The design of future things.* New York: Basic Books.

O'Connor, M. C. (2015). ARIBO: Bringing autonomous vehicles to military bases, campuses and even city streets. *IOT Journal.* Retrieved from www.iotjournal.com/articles/view?12607.

OpenPlans. (2012). OpenTripPlanner. opentripplanner.com (Accessed August 16, 2012).

Parker, D. J. (2008). *Transit Cooperative Research Program Synthesis 73: AVL systems for bus transit: Update.* Washington, DC: National Academy Press.

PAVIP. (2009). *Project brochure: PAVIP transport.* Your Bones. Retrieved from http://bones.ch/bones/media/downloads-eng/pavip/Flyer%20PAVIP%20Transport_English.pdf (Accessed August 16, 2012).

Perez, B., Kocher, L., Nemade, M., Paquet, V., & Lenker, J. (2016). *Usability of transit bus equipment for access and securement for individuals with mobility impairment.* AHFE 2016 International Conferences in Orlando, FL, July 27–31, 2016.

Petrella, M., Rainville, L., & Spiller, D. (2009). *Remote infrared audible signage pilot/program evaluation report.* Report No. FTA-MA-26-7117-2009.01. Washington, DC: Federal Transit Administration, U.S. Department of Transportation. Retrieved from www.fta.dot.gov/documents/RIAS_EvaluationReport.pdf.

Petrik, K. (2009). Participation and e-democracy how to utilize web 2.0 for policy decision-making. In *Proceedings of the 10th Annual International Conference on Digital Government Research: Social Networks: Making Connections between Citizens, Data and Government.*

Petzall, J. (1993). Ambulant disabled persons using buses: Experiments with entrances and seats. *Applied Ergonomics, 24*(5), 313–326.

Prahalad, C. K., & Ramaswamy, V. (2000). Co-opting customer competence. *Harvard Business Review, 78*(1), 79–90.

Project for Public Spaces. (1997). *The role of transit in creating livable metropolitan communities, Transit Cooperative Research Program, Report 22.* National Academy Press. Retrieved from http://onlinepubs.trb.org/onlinepubs/tcrp/tcrp_rpt_22-a.pdf.

Q'Straint. (2012). *Q'Straint quantum securement system.* Retrieved on November 5, 2012 from www.qstraint.com/en_na/products/transit-solutions/quantum.

Qian, H. (2010). Global perspectives on e-governance. In *Proceedings of the 4th International Conference on Theory and Practice of Electronic Governance.*

Repenning, A., & Ioannidou, A. (2006). Mobility agents: Guiding and tracking public transportation users. In *Proceedings from International Working Conference on Advanced Visual Interfaces.*

RERC-WTS. (2012a). Rehabilitation Engineering Research Center on Wheelchair Transportation Safety. Last updated May 28, 2009. Retrieved October 2, 2012, from www.rercwts.org.

RERC-WTS. (2012b). *The WC19 information resource: Crash-tested wheelchairs & seating systems.* Rehabilitation Engineering Research Center on Wheelchair Transportation Safety. Last updated August 18, 2010. Retrieved October 2, from www.rercwts.org/wc19.html.

Reuters. (2011, April 26). *Commuting impacts job satisfaction-poll.* Retrieved from www.reuters.com/article/2011/04/26/uk-work-commuting-health-idUSLNE73P02H20110426.

Richmond, G., & Steinfeld, E. (1999). The usability of tactile warning surfaces for people with visual impairments. In E. Steinfeld & G. S. Danford (Eds.), *Measuring enabling environments*. New York: Kluwer Science/Plenum.

Ritter, A., Straight, A., & Evans, E. (2002). *Understanding senior transportation*. Washington, DC: AARP Public Policy Institute.

Roberts, P., & Babinard, J. (2005). *Transport strategy to improve accessibility in developing countries*. Washington, DC: World Bank. Retrieved from http://siteresources.worldbank.org/DISABILITY/Resources/2806581172672474385/TransportStrategyRoberts.pdf.

Rodrigo, D., & Amo, P. A. (2006). *Background document on public consultation*. Paris: OECD.

RouteShout, www.routeshout.com (Accessed August 16, 2012).

RTC Rural. (1999). *Rural facts: Inequities in rural transportation*. Retrieved from http://rtc.ruralinstitute.umt.edu/Trn/TrnInequitiesFact.htm (Accessed August 30, 2012).

Rubinstein, Z. B. (2002). *Efficient scheduling of evolving, nondeterministic process plans in dynamic environments*. PhD thesis, University of Massachusetts, Amherst, MA, August 2002.

Rubinstein, Z. B., & Smith, S. F. (2011, June). Dynamic management of paratransit vehicle schedules. In *Proceedings of the 5th International Workshop on Scheduling and Planning Applications (SPARK-2011)*, Freiburg, Germany.

Rubinstein, Z. B., Smith, S. F., & Barbulescu, L. (2012, July). Incremental management of oversubscribed vehicle schedules in dynamic dial-a-ride problems. In *Proceedings of the 26th Annual Conference of the Association for the Advancement of Artificial Intelligence (AAAI 2012)*, Toronto, CA.

Russell, L. (1999). *The future of the built environment*. The Millennium Papers. London: Age Concern England.

Rutenberg, U. (1995). *Urban transit bus accessibility considerations*. Rutenberg Design Incorporated. Toronto, ON: Canadian Urban Transit Association, 110 pp.

Rutenberg, U., & Hemily, B. (2003). *Use of rear-facing position for common wheelchairs on transit buses*. Transit Cooperative Research Program (TCRP) Synthesis 50. Washington, DC: Transportation Research Board.

Saad-Sulonen, J., & Cabrera, A. B. (2008). Setting up a public participation project using the urban mediator tool: A case of collaboration between designers and city planners. In *Proceedings of NordiCHI*. ACM Press, pp. 539–542.

Saad-Sulonen, J., Botero, A., & Kuutti, K. (2012). A long-term strategy for designing (in) the wild: Lessons from the urban mediator and traffic planning in Helsinki. In *Proceedings of DIS*. ACM Press, pp. 166–175.

Said, C. (2015). *Deaf drivers flocking to Lyft*. SFGate. www.sfgate.com/bayarea/article/Deaf-drivers-flocking-to-Lyft-5989305.php.

Santos, A., McGuckin, N., Nakamoto, H. Y., Gray, D., & Liss, S. (2011). *Summary of travel trends: 2009 National Household Travel Survey*. FHWA-PL-11-022. Washington, DC: U.S. DOT, FHWA.

Sato, D., Takagi, H., Kobayashi, M., Kawanaka, S., & Asakawa, C. (2010). Exploratory analysis of collaborative web accessibility improvement. *ACM Transactions on Accessible Computing, 3*(2), article 5. Retrieved from http://doi.acm.org/10.1145/1857920.1857922.

Schneider, L. W., Manary, M. A., Hobson, D. A., & Bertocci, G. E. (2008). Transportation safety standards for wheelchair users: A review of voluntary standards for improved safety, usability, and independence of wheelchair-seated travelers. *Assistive technology, 20*(4), 222–233.

Schneider, W., & Brechbuhl, A. (1991). *Defining the low-floor bus: Its advantages and disadvantages*. Paper presented at the 49th UITP Congress, Stockholm.

Schwartz, J. D. (2011). 25% of motorists hate their commute. *The Urban Country*. Retrieved from www.theurbancountry.com/2011/01/25-of-motorists-hate-their-commute.html.

Schwartzel, E. (2010). Port Authority takes to Twitter to report problems. *Pittsburgh Post-Gazette*.

Schweiger, C. (2003). *Real-time bus arrival information systems*. TCRP Synthesis 48. Washington, DC: Transportation Research Board.

Schweitzer, L. (2011). Public transit's imperiled future. *Progressive Planning, 189*, 1–7.

Silvertown, J. (2009). A new dawn for citizen science. *Trends in Ecology & Evolution, 24*(9), 467–471.

Skedaddle. (2015). From www.letskedaddle.com/.

Smith, A. (2015). *U.S. smartphone use in 2015*. Pew Research Center. Retrieved from www.pewinternet.org/2015/04/01/us-smartphone-use-in-2015/.

Smith, S. F., Becker, M. A., & Kramer, L. (2004). Continuous management of airlift and tanker resources: A constraint-based approach. *Mathematical and Computer Modeling, 39*(6–8), 581–598.

Southworth, M. (2005). Designing the walkable city. *Journal of Urban Planning and Development, 131*, 246–257.

Spas, D., & Seekins, T. (1998). *Rural facts: Rural transportation*. RTC: Rural. Retrieved from http://rtc.ruralinstitute.umt.edu/Trn/TrnFact.htm.

Steinfeld, A., Rao, S. L., Tran, A., Zimmerman, J., & Tomasic, A. (2012). Co-producing value through public transit information services, International Conference on Human Side of Service Engineering (co-located with the International Conference on Applied Human Factors and Ergonomics). Retrieved from www.ahfe2012.org/.

Steinfeld, A., Zimmerman, J., Tomasic, A., Yoo, A., & Aziz, R. (2012). Mobile transit rider information via universal design and crowdsourcing. *Transportation Research Record-Journal of the Transportation Research Board, 2217*, 95–102.

Steinfeld, E. (2001). Universal design in mass transportation. In W. Preiser & E. Ostroff (Eds.), *Universal design handbook*. New York: McGraw Hill, pp. 24.23–24.25.

Steinfeld, E. (2010). Advancing universal design. In J. L. Maisel (Ed.), *The state of the science in universal design: Emerging research and developments*. Oak Park, IL: Bentham Science Publishers, pp. 1–19.

Steinfeld, E., D'Souza, C., & Maisel, J. (2010). *Clear floor space for contemporary wheeled mobility users*. Paper presented at the 12th International Conference on Mobility and Transport for Elderly and Disabled Persons (TRANSED 2010), Hong Kong, June 2–4, 2010.

Steinfeld, E., Grimble, M., Steinfeld, A., Rao, S. L., & Tran, A. (2012). *Identifying accessibility problems in existing transit systems*. Paper presented at the 13th International Conference on Mobility and Transport for Elderly and Disabled Persons (TRANSED 2012), New Delhi, India, September 17–21, 2012.

Steinfeld, E., Grimble, M., & White, J. (2011). *Guided Tours field study of public transportation*. Poster Presentation. FICCDAT 2011, Toronto, ON.

Steinfeld, E., & Maisel, J. L. (Eds.). (2012). *Universal design: Creating inclusive environments*. Hoboken, NJ: Wiley & Sons.

Steinfeld, E., Maisel, J., & Feathers, D. (2005). *Standards and anthropometry for wheeled mobility for U.S. Access Board*. Buffalo, NY: IDEA Center.

Steinfeld, E., Maisel, J., Feathers, D., & D'Souza, C. (2010a). Anthropometry and standards for wheeled mobility: An international comparison. *Assistive Technology, 22*(1), 51–67.

Steinfeld, E., Paquet, V., D'Souza, C., Joseph, C., & Maisel, J. (2010b). *Anthropometry of wheeled mobility project: Final report*. Buffalo, NY: Prepared for U.S. Access Board by the IDeA Center.

Steinfeld, E., & Seelman, K. (2011). Enabling environments. In *World Report on Disability*. Geneva: World Health Organization.

Stineman, M. G., Ross, R. N., Fiedler, R., Granger, C. V., & Maislin, G. (2003). Functional independence staging: Conceptual foundation, face validity, and empirical derivation. *Archives of Physical Medicine and Rehabilitation, 84*(1), 29–37.

Subryan, H., Landau, S., & Steinfeld, E. (2012). *Universal design interactive multisensory models*. The 4th International Conference for Universal Design, Fukuoka, Japan, October 12–14, 2012.

Talking Signs. (2008). Retrieved from www.talkingsigns.com (Accessed August 16, 2012).

Tay, R. (2006). Ageing drivers: Storm in a teacup? *Accident Analysis & Prevention, 38*(1), 112–121.

Thiagarajan, A., Ravindranath, L., LaCurts, K., Madden, S., Balakrishnan, H., Toledo, S., & Eriksson, J. (2009). VTrack: Accurate, energy-aware road traffic delay estimation using mobile phones. In *Proceedings of the ACM Conference on Embedded Networked Sensor Systems (SenSys)*.

Tomasic, A., Zimmerman, J., Garrod, C., Huang, Y., Nip, T., & Steinfeld, A. (2015). *The performance of a crowdsourced transportation information system*. Transportation Research Board 2015 Annual Meeting. Washington, DC: Transportation Research Board.

Tomkins, A. J., Pytlik Zillig, L. M., Herian, M. N., Abdel-Monem, T., & Hamm, J. A. (2010). Public input for municipal policymaking. In *Proceedings of the 11th Annual International Digital Government Research Conference on Public Administration Online: Challenges and Opportunities*.

Touch Graphics. (2008). New York City's Growing Network of Talking Kiosks, Access and the City Conference, Dublin, Ireland 2008.

Trace Center. (2000). *Making Information/Transaction Machines (ITMs) accessible.* Final Report, Spring 2000. Retrieved from http://trace.wisc.edu/world/kiosks/itms/index.html (Accessed August 16, 2012).

TransitCare Ltd. (2015). *About Us.* Retrieved from www.transitcare.com.au/about-us/.

Transportation Research Board. (2012). *TCRP Synthesis 96: Off-board fare payment using proof-of-payment verification.* Washington, DC: Transportation Research Board.

Turkovich, M., van Roosmalen, L., Hobson, D., & Porach, E. (2011). Alternative wheelchair securement systems—Performance during normal and emergency driving in a public bus. *Public Transportation Journal, 14*(3), 147–169.

Turner, D. S., Evans, W. A., Wolshon, B., Dixit, V., Sisiopiku, V. P., Islam, S., & Anderson, M. D. (2010). *Transportation-oriented communication with vulnerable populations during major emergencies.* Transportation Research Board 2010 Annual Meeting.

Uber Newsroom. (2015). *Announcing: Ubermilitary families coalition & Ubermilitary director.* Uber. Retrieved from https://newsroom.uber.com/2015/05/announcing-ubermilitary-families-coalition-ubermilitary-director/.

ULI (Annual Reports). (2009). *Emerging trends in real estate.* Urban Land Institute. Retrieved from www.uli.org/ResearchAndPublications/EmergingTrends/Americas.aspx.

University of Michigan Transportation Research Institute (UMTRI). (2011, October–December). Driving forces. Fewer young, but more elderly, have driver's licenses. *Research Review, 42*(4), 1–2.

U.S. Access Board. (2009). *Recovery requires accessibility.* Retrieved from www.access-board.gov/recovery/.

U.S. Access Board. (2016). *Americans with Disabilities Act (ADA) accessibility guidelines for transportation vehicles.* Federal Register (36 CFR Part 1192). Retrieved from https://access-board.gov/guidelines-and-standards/transportation/vehicles/update-of-the-guidelines-for-transportation-vehicles/final-updated-guidelines-for-buses-and-vans/single-file-version.

U.S. Access Board and Department of Transportation. (1998). *ADA accessibility guidelines for transportation vehicles.* Federal Register (36 CFR Part 1192). Retrieved from www.access-board.gov/transit/html/vguide.htm.

U.S. Access Board and Department of Transportation. (2007a). *Draft revisions to the ADA accessibility guidelines for buses and vans.* Retrieved July 22, 2008, from www.access-board.gov/vguidedraft.htm.

U.S. Access Board and Department of Transportation. (2007b). *Public comments on the draft update of guidelines for buses and vans.* Retrieved July 22, 2008, from www.access-board.gov/transit/comments/.

U.S. Census Bureau. (2001). *American Housing Survey.* Retrieved October 5, 2012, www.census.gov/hhes/www/housing/ahs/ahs01_2000wts/ahs01_2000wts.html.

U.S. Census Bureau. (2006). 2006 *American Community Survey.*

U.S. Dept of Justice, Civil Rights Division, & United States of America. (2004). Enforcing the ADA: A Status Report From the Department of Justice.

U.S. Department of Transportation. (1988). *49 C.F.R Section 604.3(g): Circular 2710.2A.*

U.S. Department of Transportation. (2007). Federal Register 49 CFR Part 38. Americans with Disabilities Act (ADA) Accessibility Specifications for Transportation Vehicles (Revised as of October 1, 2007). Retrieved February 17, 2010, from www.fta.dot.gov/civilrights/ada/civil_rights_3905.html.

U.S. Department of Transportation. (2016). Architectural Barriers and Transportation Board, "Americans with Disabilities Act (ADA) Accessibility Guidelines for Transportation Vehicles", Federal Register, Vol. 81, No. 240, pp. 90600–90629, 12/14/16, 36 CFR Part 1192 [Docket No. ATBCB 2010-004].

van Roosmalen, L., Karg, P., Hobson, D., Turkovich, M., & Porach, E. (2011). User evaluation of three wheelchair securement systems in large accessible transit vehicles. *Journal of Rehabilitation Research and Development, 48*(7), 823–838.

Vanderheiden, G. C. (1997). Cross disability access to touch screen kiosks and ATMs. *Advances in Human Factors/Ergonomics, 21A*, 417–420.

Vanderheiden, G. C., Law, C. M. & Kelso, D. (1999, May). *EZ access interface techniques for anytime anywhere anyone interfaces.* Pittsburgh, PA: CHI'99 (Computer-Human Interaction).

Venter, C., Mashiri, M., Rickert, T., Maunder, D., Sentinella, J., de Deus, K., …, Bogopane, H. (n.d.). *Towards the development of comprehensive guidelines for practitioners in developing countries.* United Kingdom: Department for International Development (DFID).

Verifone Transportation Systems. (2016). About Curb. Retrieved from https://gocurb.com/about/.

VETaxi. www.vetaxi.com (Accessed October 28, 2016).

Von Ahn, L., Maurer, B., McMillen, C., Abraham, D., & Blum, M. (2008). Recaptcha. Human-based character recognition via web security measures. *Science, 321*, 5895.

Wahl, H., Oswald, F., & D. Zimprich. (1999). Everyday competence in visually impaired older adults: A case for person-environment perspectives. *The Gerontologist, 39*(2), 140–149.

Web Accessibility Initiative. (WAI). Retrieved from www.w3.org/WAI (Accessed August 16, 2012).

Weingroff, R. M. (2011, April 7). When did the Federal Government begin collecting the gas tax? *Ask the Rambler.* Washington, DC: Federal Highway Administration.

Weir, K. (2009). Blind man falls onto Metro tracks, isn't the first to slip. *The Washington Examiner.* Retrieved from http://washingtonexaminer.com/local/blind-man-falls-metro-tracks-isn039t-first-slip.

Welch, P. (1995). What is universal design? In P. Welch (Ed.), *Strategies for teaching universal design.* Boston, MA: Adaptive Environments Center and MIG Communications.

Winters, P. L., Barbeau, S. J., & Georggi, N. L. (2010). *Travel Assistance Device (TAD) to help transit riders.* Transit IDEA Project 52. Transportation Research Board. Retrieved from http://pubsindex.trb.org/view.aspx?id=923659.

Wretstrand, A., Stahl, A., & Petzall, J. (2008). Wheelchair users and public transit: Eliciting ascriptions of comfort and safety. *Technology and Disability, 20*(1), 37–48.

Wright, L. (2004). Bus rapid transit planning guide. *Division 44 Environment and Infrastructure Sector Project "Transport Policy Advice".* GTZ: Federal Ministry for Economic Cooperation and Development.

Yoo, D., Zimmerman, J., & Hirsch, T. (2013). Probing bus stop for insights on transit co-design. In *Proceedings of the Conference on Human Factors in Computing Systems (CHI).*

Yoo, D., Zimmerman, J., Steinfeld, A., & Tomasic, A. (2010). Understanding the space for co-design in riders' interactions with a transit service. In *Proceedings of the Conference on Human Factors in Computing Systems (CHI).*

Young, C. (2015). Partnership between MBTA's the ride and Uber coming for paratransit passengers. Retrieved from www.wbur.org/2015/11/10/the-ride-uber-partnership.

Zhuang, Z., Kim, K. H., & Singh, J. P. (2010). Improving energy efficiency of location sensing on smartphones. In *Proceedings of the ACM International Conference on Mobile Systems, Applications, and Services (MobiSys).*

Zimmerman, J., Tomasic, A., Garrod, C., Yoo, D., Hiruncharoenvate, C., Aziz, R., …, Steinfeld, A. (2011). Field trial of Tiramisu: Crowd-sourcing bus arrival times to spur co-design. In *Proceedings of the Conference on Human Factors in Computing Systems (CHI).*

Index

1/4 mile rule 24
3-D Printed Car 78

Access Exchange 29
accessibility 25–7; built environment xv
accessibility laws 25
accessibility services 9–10
accessible public transportation 25–7;
    challenges to 115–18; future of 118–19;
    importance of 1–2; priorities for 120–1
AccessMyNYC 95–6
active transportation 8
ADA (Americans with Disabilities Act) 33;
    accessible information 35; bus shelters 42;
    shared cars 76; taxis 72; TNCs (transportation
    network companies) 73; vehicle standards 13;
    wheelchair securement 64
ADAAG (Americans with Disabilities Act
    Accessibility Guidelines) 57
ADA/Section 504 Coordinator 9
advertising agencies, bus shelters 10–11
affordability, fare reductions 22–3
Airport Flyer 11
airport parking 11
Alamo Rent-A-Car LLC 74
alt text 34
AM General 71
Amazon 94
American Public Transportation Association
    (APTA) 119
Americans with Disabilities Act (ADA) 33;
    accessible information 35; bus shelters 42;
    shared cars 76; taxis 72; TNCs (transportation
    network companies) 73; vehicle standards 13;
    wheelchair securement 64
Americans with Disabilities Act Accessibility
    Guidelines (ADAAG) 57
ANC Rental Corporation 74
APC (automatic passenger counting) 61
applicability, paratransit 88
apps, smartphone apps 36–7
APTA (American Public Transportation
    Association) 119
attracting riders 26
automatic passenger counting (APC) 61

automatic vehicle location (AVL) 3, 87
automobile transportation 6–7
automobiles: autonomous vehicles 77–9; design
    of 15–16 see also vehicle design; digital
    modeling 18
autonomous vehicles 77–9
Avis 76
AVL (automatic vehicle location) 3, 87

barriers 52
BART (Bay Area Rapid Transit) 46, 116
B/D (boarding/disembarking) 54, 115; floor
    plans 59–60
best practices, identifying 29–31
bike sharing programs 8
boarding/disembarking (B/D) 54, 115
Bridj 75
broken equipment, lifts/elevators 25–6
BRT (Bus Rapid Transit) 29, 30
built environment 39–51
bulb-outs 58
Bus Rapid Transit (BRT) 29, 30
bus shelters 42–3; franchise-bid
    programs 10–11
buses 54–5; broken equipment 25–6; design
    of 15–16; floor plans 59–60; high-floor
    buses 55–6; low-floor bus xiv, 56; wheelchair
    securement systems 64–5

Cache 93
call centers 33–4
Carma 76
case studies, Tiramisu 102–4
cash-based payment systems 61
cell tower localization 93
challenges to accessible public transportation
    115–18
changing platform levels 50–1
citizen participation 100
citizen science 99
CityMobil2 project 78
co-design 102
co-designers, riders as 107–13
cognitive maps 92
communication services 116

Commuter Connections 76–7
connected vehicle tracking systems, paratransit 87
contact-based cards, electronic fare payment systems 61
contactless proximity cards 61
crowdsourcing 99
Curb 72

DC Metro (Washington Metropolitan Area Transportation Authority) 9, 10; web-based information 34
demand responsive transportation *see* DRT (demand responsive transportation)
design, service design 101–2
design of: buses 15–16; vehicle design *see* vehicle design
destination verification 91
diaries, capturing rider data 112–13
digital information 90–1
digital information technologies xv–xvi
digital modeling, automobiles 18
directional RIAS 91
disabilities, travel patterns of people with disabilities 22
disability rates 21
DOT (U.S. Department of Transportation) 13–15
driverless cars 77–9
driving fatalities 8
driving participation 8
DRT (demand responsive transportation) 68; autonomous vehicles 77–9; defined 68–9; partnership models 79–80; payment technologies 80–1; sharing cars and rides 76–7; shuttle buses 74–6; STS (special transport services) 69–71; taxis, hired vehicles, and transportation network companies 71–3
dynamic scheduling 87–8
dynamic signs, stations and terminals 44

Easy Mile 78
economic benefits of accessible public transportation 2
eGovernance 100
egress: level changes, vehicle design 55–8; vehicle design 61
elderly riders 117
electronic fare payment systems 61
electronic infrastructure 91–2
elevators 50
energy policy 118
entry, vehicle design 61
EZ10, Local Motors 78

fare gates 23
fare payment, vehicle design 61–2
fare reductions 22–3

fare systems 47
fatalities, driving 8
first mile problem 24
fixed route systems 25
FixMyStreet.com 100
flex routes 25
Flexcar 76
floor plans, vehicle design 59–60
foreign languages 46–7
Forward Entry-Forward Exit, buses 59–60
franchise-bid programs 10–11
fullness data 3
future of public transportation 118–19

gaps, level changes 50
Geo-Access Challenge Team 95, 116
geolocation data, integrating with information about accessible infrastructure 95–6
Global Positioning System (GPS) 45, 93; paratransit 87
Goals of Universal Design 28
Google 94
Google Car 77
government, social computing and 99–100
GPS (Global Positioning System) 45, 93; paratransit 87
guaranteed ride home feature, Commuter Connections 77
guide paths 52
guide strips 45
Guided Tours 108–9

help lines 33–4
high-capacity highway model 6
high-floor buses 55–6
hired vehicles 71–3
horizontal gap problem 55
human computation 99
human digital models 18

IBM: AccessMyNYC 95–6; Social Accessibility Project 35
identifying best practices 29–31
ideograms 44–5
IDRM (International Disability Rights Monitor) 29
importance of public transportation 1–2
inclusive public transportation xiv, 21–4
Indian Trails 36
inferring location 92–4
information, situational awareness 2–3
information kiosks 47
infrastructure-based information, trip planning and rider information 35–6
ingress, level changes, vehicle design 55–8
integrated accessibility information 95–6

International Conference on Mobility and Transport for Elderly and Disabled Persons (TRANSED) 13
International Disability Rights Monitor (IDRM) 29

jitney service 74

knowledge models, applying need to 3–4

LA Metro (Los Angeles County Metropolitan Atlanta Regional Transportation Authority) 9, 10, 11; Orange Line 40
"last mile" problem xiv, 24, 121
Lausanne Metro 36
layout, stations and terminals 43–4
Let's Go! project 34
level changes: built environment 48–9; vehicle design 55–8; vehicle loading 49–50
lifts 50, 55, 56
live schedules 87
loading platforms, safety issues 51–3
localization 91–2
location-based information 90–5, 116
locations, inferring 92–4
Los Angeles County Metropolitan Transportation Authority (LA Metro) 9, 10, 11; Orange Line 40
low-density communities, strategies for serving 115
low-floor bus xiv, 28–9, 56
Lusher, Ruth Hall 27
Lyft 69, 73, 79, 80

Mace, Ron 27
machine learning, location-based information 94–5
Mahan, Steve 77
Markov model 93
MARTA (Metropolitan Atlanta Regional Transportation Authority) 9
MBTA (Massachusetts Bay Transportation Authority) 9, 10
message boards 47
Metropolitan Atlanta Regional Transportation Authority (MARTA) 9
metropolitan transportation agencies, policies and regulations 13–15
Mid Entry-Forward Exit, buses 59–60
Mid Entry-Mid Exit configuration, buses 59–60
mid-boarding 61
mini-platforms 51
Multisensory Interactive Touch Model 48
MV1 71–2

National Car Rental System, Inc. 74
National Institute on Disability, Independent Living, and Rehabilitation Research (NIDILRR) 9

navigation 90–1
NavPal 37
Near Field Communication (NFC) 36
New York City Taxi and Limousine Commission 71–2
NFC (Near Field Communication) 36
NFC tags 93–4
NFTA (Niagara Frontier Transportation Authority) 9
NIDILRR (National Institute on Disability, Independent Living, and Rehabilitation Research) 9
no show policies, STS (special transport services) 70–1

obesity epidemic 117
older drivers 2
older riders 117
on-demand service 25
on-demand transportation xv
OneBusAway 37
one-time trips, paratransit 85
online surveys 110–12
OpenTripPlanner 37
option generation, dynamic scheduling 88
Orange Line, LA Metro (Los Angeles County Metropolitan Atlanta Regional Transportation Authority) 40
orientation 90

PAAC (Port Authority of Allegheny County) 9, 99, 110
paratransit xiii, 12, 25, 26, 69–71, 83–4; broader applicability 88; scheduling and routing problems 84–6; technology for scheduling 86–8
ParkScan 100
participatory sensing 99
partnership models 79–80
payment technologies 80–1
PBD (programming-by-demonstration) 95
pedestrian environment 24, 40, 115
people with disabilities, travel patterns 22
pickup delay constraints 84
pickup window constraints 84
platform edges 51
platform levels, changing 50–1
platform loading 50–1
platforms, safety issues 51–3
point-of-pain experiences 104
policies, transportation service and policies 21–4
policies and regulations 13–15
Port Authority of Allegheny County (PAAC) 9, 99, 110
Port Authority of LA County 11
poverty rates 21
Principles of Universal Design 27–9
printed media, trip planning and rider information 33–4

priorities for accessible public transportation 120–1
programming-by-demonstration (PBD) 95
psychological travel time 99
public address announcements 47
public agencies 9–13
public consultation 100
public services 101
public transportation, importance of 1–2
public transportation agency practices 9–13

Q-Pod 65
QR codes 93–4
Quantum 65

ramps 48, 50; vehicle design 56–8
Real Time Reporting 109–10
real-time arrival data systems 33–4
real-time arrival estimates 3
real-time arrival information, OneBusAway 37
real-time information, providing 116
Rear Entry-Front Exit, buses 59
Rear Entry-Rear Exit, buses 59
regional development 6–8
regulations 13–15
Rehabilitation Engineering Research Center on Accessible Public Transportation (RERC-APT) 9
remote human help on demand 47
Remote Infrared Audible Signs (RIAS) 45–6
RERC program, simulation methods 18
RERC-APT (Rehabilitation Engineering Research Center on Accessible Public Transportation) 9
rescheduling, paratransit 85
research, policies and regulations 14
reservation systems, STS (special transport services) 70–1
RFID tags 93–4
RF-WPS 65
RIAS (Remote Infrared Audible Signs) 45, 91
Rickert, T. 29
rider information 32–7, 120
riders: attracting 26; as co-designers 107–13
ride-time constraints 84
robotic technology xv
rural communities 117; transportation service and policies 21–2

safety issues, platforms 51–3
scheduling and routing problems, paratransit 84–8
self-driving cars 77–9
SensLoc 93
SEPTA (Southeast Pennsylvania Transportation Authority) 9, 10, 12, 25, 48, 50
service delivery xiv, 107
service design xvi, 98, 101–4
service interaction 120
service quality 116

services 107
shared cars 76–7
shuttle buses 74–6
signs, stations and terminals 44–6
simulation methods 18–19
single-use zoning model 6
situational awareness 2–3
Skedaddle 76
slope of ramps 56–8
smartpens 92
smartphone apps 36–7; ride sharing 76
smartphone screen readers 104
smartphones 33, 118; localization 93; payment 81
Social Accessibility Project, IBM 35
social computing xvi, 98–104
software agents 47
Southeast Pennsylvania Transportation Authority (SEPTA) 9
special transport services (STS) 68–71
sprawl 6–7
stairs, vehicle design 55
stakeholder involvement 120
stations 43–7; level changes 48–9
STS (special transport services) 68–71; partnership models 79–80
subdivisions 6
subscription trips, paratransit 85
suburban communities 117–18
suburbs 117–18
surveys, online surveys 110–12

tactile cues 45
tactile maps 46
tactile tiles 51–2
TAD (Travel Assistance Device) 37
tags, localization 93–4
TalkBalk 104
Talking Signs 45
TARDEC ARIBO 78
taxis 25, 71–3
technology xv; payment technologies 80–1; for scheduling paratransit 86–8
telephone call centers 33–4
terminals 43–7; level changes 48–9
text messages 35–7
three-dimensional models 47
tie-down systems, wheelchair securement systems 64–5
time sharing services 69
time-sensitive information 47
Tiramisu 102–4
T-Money system 80–1
TNCs (transportation network companies) 68, 71–3; STS (special transport services) 79–80
TND (Traditional Neighborhood Development) 7
TOD (Transit Oriented Development) 115
tools: simulation methods 18–19; Transect 17; Travel Chain 17–18

traditional information channels, trip planning and rider information 33–4
Traditional Neighborhood Development (TND) 7
traffic, transit stops 41
training, public agencies 12
Transect 17
TRANSED (International Conference on Mobility and Transport for Elderly and Disabled Persons) 13, 30
transit agencies, situational awareness 2–3
Transit Oriented Development (TOD) 115
transit ridership 8
transit stops 40–3
TransitCare 80
transportation agencies, policies and regulations 13–15
transportation dependency 21
transportation network companies (TNCs) 68, 71–3; STS (special transport services) 79–80
transportation policy and regulations 13–15
Transportation Research Board (TRB) 13, 29–30
transportation service and policies, inclusive public transportation 21–4
Travel Assistance Device (TAD) 37
Travel Chain xiv, 17–18
travel information, text messages 35–6
travel patterns 22
TRB (Transportation Research Board) 13, 29–30
trip planning 32–7
trip-planning information xiv–xv
Twitter 99

Uber 69, 73, 79, 80
UberASSIST 80
UberMILITARTY 80
universal design xiv, 21, 27–9, 120
urban development 6–8
Urban Mediator 100
urban planning 99–100

U.S. Access Board 13–15; ramps 57
U.S. Department of Transportation *see* DOT (U.S. Department of Transportation)
U.S. National Council on Disability 70
vehicle design xv, 15–16, 65–7, 120; change in levels 55–8; entry/egress location 61; fare payment 61–2; floor plans 59–60; wheelchair securement spaces 62–4; wheelchair securement systems 64–5

vehicle loading 49–50
vehicles, autonomous vehicles 77–9
VETaxi 71
visibility, bus shelters 42
VoiceOver 104
VTrack 93

walkability 24
warning signals 51
Washington Metropolitan Area Transportation Authority (DC Metro) 9
WAVE 34
wayfinding xv, 90–1 *see also* location-based information
Waymo 77
WC19 standards 64
WebAim.org 34
web-based information, trip planning and rider information 34–5
wheelchair securement spaces 62–4
wheelchair securement systems 64–5
wheelchairs 14–15, 57–8; securement spaces 62–4; securement systems 64–5, 117
wheeled mobility devices, accidents 57
Wi-Fi access point maps 93
will-call requests, paratransit 86

ZipCar 69, 76
zoning patterns 6